Cooperative Path Planning of Unmanned Aerial Vehicles

Aerospace Series List

Cooperative Path Planning of Unmanned Aerial Vehicles	Tsourdos et al	November 2010
Principles of Flight for Pilots	Swatton	October 2010
Air Travel and Health: A Systems Perspective	Seabridge et al	September 2010
Design and Analysis of Composite Structures: With Applications to Aerospace Structures	Kassapoglou	September 2010
Unmanned Aircraft Systems: UAVS Design, Development and Deployment	Austin	April 2010
Introduction to Antenna Placement & Installations	Macnamara	April 2010
Principles of Flight Simulation	Allerton	October 2009
Aircraft Fuel Systems	Langton et al	May 2009
The Global Airline Industry	Belobaba	April 2009
Computational Modelling and Simulation of Aircraft and the Environment: Volume 1 – Platform Kinematics and Synthetic Environment	Diston	April 2009
Handbook of Space Technology	Ley, Wittmann Hallmann	April 2009
Aircraft Performance Theory and Practice for Pilots	Swatton	August 2008
Surrogate Modelling in Engineering Design: A Practical Guide	Forrester, Sobester, Keane	August 2008
Aircraft Systems, 3rd Edition	Moir & Seabridge	March 2008
Introduction to Aircraft Aeroelasticity And Loads	Wright & Cooper	December 2007
Stability and Control of Aircraft Systems	Langton	September 2006
Military Avionics Systems	Moir & Seabridge	February 2006
Design and Development of Aircraft Systems	Moir & Seabridge	June 2004
Aircraft Loading and Structural Layout	Howe	May 2004
Aircraft Display Systems	Jukes	December 2003
Civil Avionics Systems	Moir & Seabridge	December 2002

Cooperative Path Planning of Unmanned Aerial Vehicles

Antonios Tsourdos, Brian White and Madhavan Shanmugavel

Cranfield Defence and Security
Cranfield University, UK

A John Wiley and Sons, Ltd., Publication

Registered office
John Wiley & Sons Ltd, The Atrium, Southern Gate, Chichester, West Sussex, PO19 8SQ, United Kingdom

For details of our global editorial offices, for customer services and for information about how to apply for permission to reuse the copyright material in this book please see our website at www.wiley.com.

Library of Congress Cataloguing-in-Publication Data

Tsourdos, Antonios.
  Cooperative path planning of unmanned aerial vehicles / Antonios Tsourdos, Brian White, Madhavan Shanmugavel.
    p. cm.
  Includes bibliographical references and index.
  ISBN 978-0-470-74129-0 (cloth)
   1. Drone aircraft–Automatic control. 2. Guidance systems (Flight) 3. Airplanes–Piloting–Mathematics. 4. Airplanes–Piloting–Planning. 5. Airways–Mathematical models. I. White, Brian, 1947 June 6- II. Shanmugavel, Madhavan. III. Title.

  TL589.4.T78 2010
  629.132'5 – dc22

                                                      2010026275

A catalogue record for this book is available from the British Library.

Print ISBN: 978-0-470-74129-0
ePDF ISBN: 978-0-470-97520-6
oBook ISBN: 978-0-470-97463-6
ePub ISBN: 978-0-470-97464-3

Typeset in 10/12.5pt Palatino by Laserwords Private Limited, Chennai, India
Printed and Bound in Singapore by Markono Print Media Pte Ltd

Contents

About the Authors   ix

Series Preface   xi

Preface   xii

Acknowledgements   xiii

List of Figures   xv

List of Tables   xxi

Nomenclature   xxiii

1 Introduction   1
   1.1 Path Planning Formulation   2
   1.2 Path Planning Constraints   3
        1.2.1 Flyable Paths: Capturing Kinematics   4
        1.2.2 UAV Inertial Manoeuvre Coordinates   6
        1.2.3 Generation of Safe Paths for Path Planning   7
   1.3 Cooperative Path Planning and Mission Planning   7
   1.4 Path Planning – An Overview   10
   1.5 The Road Map Method   13
        1.5.1 Visibility Graphs   14
        1.5.2 Voronoi Diagrams   14

1.6   Probabilistic Methods   16
1.7   Potential Field   16
1.8   Cell Decomposition   17
1.9   Optimal Control   18
1.10 Optimization Techniques   18
1.11 Trajectories for Path Planning   19
1.12 Outline of the Book   20
References   22

2 **Path Planning in Two Dimensions**   **29**
2.1   Dubins Paths   30
2.2   Designing Dubins Paths using Analytical Geometry   31
      2.2.1   Dubins Path: External Tangent Solution   33
      2.2.2   Dubins Path: Internal Tangent Solution   35
2.3   Existence of Dubins Paths   37
2.4   Length of Dubins Path   39
2.5   Design of Dubins Paths using Principles of Differential
      Geometry   39
      2.5.1   Dubins Path Length   43
2.6   Paths of Continuous Curvature   45
2.7   Producing Flyable Clothoid Paths   46
2.8   Producing Flyable Pythagorean Hodograph Paths (2D)   56
      2.8.1   Design of Flyable Path using 2D PH curve   61
References   62

3 **Path Planning in Three Dimensions**   **65**
3.1   Dubins Paths in Three Dimensions Using Differential
      Geometry   67
3.2   Path Length–Dubins 3D   72
3.3   Pythagorean Hodograph Paths–3D   72
      3.3.1   Spatial PH Curves   73
3.4   Design of Flyable Paths Using PH Curves   74
      3.4.1   Design of Flyable Paths   75
References   78

4 **Collision Avoidance**   **81**
4.1   Research into Obstacle Avoidance   83
4.2   Obstacle Avoidance for Mapped Obstacles   85
      4.2.1   Line Intersection Detection   86

		4.2.2	Line Segment Intersection	90
		4.2.3	Arc Intersection	94
	4.3	Obstacle Avoidance of Unmapped Static Obstacles		103
		4.3.1	Safety Circle Algorithm	104
		4.3.2	Intermediate Waypoint Algorithm	104
	4.4	Algorithmic Implementation		106
		4.4.1	Dubins Path Modification	107
		4.4.2	Clothoid Path Modification	107
		4.4.3	PH Path Modification	110
		4.4.4	Obstacle Avoidance in 3D	112
	References			115
5	**Path-Following Guidance**			**119**
	5.1	Path Following the Dubins Path		120
	5.2	Linear Guidance Algorithm		124
	5.3	Nonlinear Dynamic Inversion Guidance		126
	5.4	Dynamic Obstacle Avoidance Guidance		132
		5.4.1	UAV Direction Control	135
		5.4.2	Multiple Conflict Resolution	142
	References			145
6	**Path Planning for Multiple UAVs**			**147**
	6.1	Problem Formulation		149
	6.2	Simultaneous Arrival		151
	6.3	Phase I: Producing Flyable Paths		152
	6.4	Phase II: Producing Feasible Paths		152
		6.4.1	Minimum Separation Distance	153
		6.4.2	Non-Intersection Paths	154
		6.4.3	Offset Curves	155
	6.5	Phase III: Equalizing Path Lengths		156
	6.6	Multiple Path Algorithm		156
	6.7	Algorithm Application for Multiple UAVs		157
		6.7.1	2D Dubins Paths	157
		6.7.2	2D Clothoid Paths	160
	6.8	2D Pythagorean Hodograph Paths		162
	6.9	3D Dubins Paths		165
	6.10	3D Pythagorean Hodograph Paths		169
	References			174

Appendix A   Differential Geometry   **175**
            A.1   Frenet–Serret Equations   177
            A.2   Importance of Curvature and Torsion   178
            A.3   Motion and Frames   179
            References   181

Appendix B   Pythagorean Hodograph   **183**
            B.1   Pythagorean Hodograph   184
            References   185

Index   **187**

About the Authors

Antonios Tsourdos was appointed Head of the Autonomous Systems Group at Cranfield University in 2007. He obtained a PhD on Nonlinear Robust Flight Control Design and Analysis from Cranfield University in 1999. He was a member of the Team Stellar, the winning team for the UK MoD Grand Challenge (2008). He has served as an Editorial Board Member for the *Proceedings of the Institution of Mechanical Engineers, Part G: Journal of Aerospace Engineering*, the *International Journal of Systems Science*, the *IEEE Transactions on Instrumentation and Measurement*, the *International Journal on Advances in Intelligent Systems* and the international journal *Mathematics in Engineering, Science and Aerospace* (MESA). He is a member of the A|D|S Autonomous Systems Strategy Group and the A&D KTN National Technical Committee on Autonomous Systems. He is also a member of the IFAC Technical Committee on Aerospace Control, the IFAC Technical Committee on Networked Systems, the AIAA Technical Committee on Guidance, Control & Navigation, the IEEE Control System Society Technical Committee on Aerospace Control (TCAC) and the IEEE Technical Committee on Aerial Robotics and Unmanned Aerial Vehicles.

Brian White obtained a BSc degree in Engineering from the University of Leicester, followed by an MSc and PhD from UMIST, Manchester, in 1974. He has worked in the aerospace industry for the company MBDA, working on novel guidance techniques for missile systems, including one of the first developments of Kalman filters for integration into guidance systems. He has since been a faculty member of the School of Engineering at the University of Bath and at Cranfield University, where for many years he was Head of the Department of Aerospace, Power and Sensors. While in the post, he built up a strong department in Aerospace Systems and also led the Control and Guidance Group, which became a world-leading group in control, navigation and guidance for unmanned aerial vehicles (UAVs). In 2008, he retired, to become Emeritus Professor, retaining a strong interest in research for Autonomy in UAV Systems with the Autonomous Systems Group. He has published well over 100 papers in guidance, control estimation and autonomy, giving invited and keynote lectures at Universities and International Conferences. By gaining research expertise in these multiple domains, he has developed an approach that produces simple, robust and effective approaches to the design of multiple UAV systems. This book expresses some of that wide experience.

Madhavan Shanmugavel is currently a Research Fellow at Cranfield Security and Defence, Defence College of Management and Technology, Cranfield University, Defence Academy of the UK, Shrivenham. He began his research career with Cranfield University in 2007, developing path planning algorithm for unmanned vehicles. He graduated from the Indian Institute of Technology, Chennai, with a Masters Degree in 2000, and gained industrial research experience with TATA Motors, India, since 2004. In his current position, he contributed his knowledge in various projects by extending the path planning algorithms to unmanned ground and aerial vehicles. He has published 14 conference papers, and two journal papers. His research interests include cooperative systems, guidance and control, robotics and path planning of unmanned systems.

Series Preface

The Aerospace Series is a source of aviation and space technology knowledge for business professionals, industry operators and users. The Series topics covered span the design, development, manufacture, operation and support as well as infrastructure operations and developments in research and technology. Authors are drawn from within the aerospace industry as well as from universities and learning institutions from around the world.

The use of Unmanned Aerial Vehicles (UAV's) in modern military conflicts has grown exponentially over the past twenty years and as the autonomous UAV has replaced the human operator in a fast growing number of military flight operations the need for effective and efficient mission management of UAV's has become critical. One key area of developing technology associated with this requirement is "Path Planning" which is the subject addressed by this book. Here the authors Antonios Tsourdos, Brian White and Madhavan Shanmugavel provide a thorough and complete treatment of the subject from its inception in ground-based robotics to the path planning needs of the modern UAV covering kinematic and environmental constraints, safety and mission/path coordination, 2-D and 3-D path planning techniques. Sense and avoid considerations are also included. Cooperative Path Planning of Unmanned Aerial Vehicles promises to be an important reference for practitioners in the growing field of autonomous air vehicles.

Allan Seabridge, Roy Langton, Jonathan Cooper and Peter Belobaba

Preface

Path planning is a complex problem, which involves meeting the physical constraints of the unmanned aerial vehicles (UAVs), constraints from the operating environment and other operational requirements. The foremost constraint to be met is that the paths must be flyable. Flyable paths are those that meet the kinematic constraints of the UAV. Satisfying this constraint ensures that the motion of the UAV stays within the maximum bounds on manoeuvre curvature. The safety of the path is measured by the ability of the path to avoid threats, obstacles and other UAVs. The path must maintain collision avoidance with other friendly UAVs and also must be flexible enough to avoid environmental obstacles and threats. Also, additional constraints – such as generating shortest paths, and minimum fuel and energy consumption paths – can be included for better performance and efficiency of the mission.

This book has grown out of the research work of the authors in the area of path planning, collision avoidance and path following for single and multiple unmanned vehicles in the past ten years. The algorithms described here result in the planning of paths that are not only flyable and safe but also implementable for real-time applications.

Antonios Tsourdos
Brian White
Madhavan Shanmugavel

Acknowledgements

We would like to thank our colleagues who helped us to develop the path planning algorithms described in this book. In particular, we would like to thank Hyo Sang Shin, Luigi Caravita, Matt Robb, Seung Keun Kim, Samuel Lazarus and Arpita Sinha.

List of Figures

1.1	A block diagram approach to path planning	4
1.2	Autopilot and guidance control loops	5
1.3	Curvature and torsion	6
1.4	Hierarchy of mission planning	9
1.5	Existing approach to path planning	11
1.6	The road map method	13
1.7	Visibility graph	14
1.8	Voronoi diagram: polygonal fences around obstacles	15
1.9	Cell decomposition	17
2.1	CLC and CCC types of Dubins path	31
2.2	Tangent circles	32
2.3	Dubins path with external tangent	32
2.4	Dubins path with internal tangent	33
2.5	Block diagram of path planner to generate the shortest flyable paths	37
2.6	Dubins paths with ϕ_f as a free variable. The start turn is either clockwise or anticlockwise. Four possible turns on each tangent circle produce eight paths	38
2.7	Dubins arc geometry	40
2.8	Set of Dubins paths over a range of κ	44
2.9	Set of Dubins path lengths over a range of κ	44
2.10	Dubins path	46
2.11	Curvature profiles of Dubins and clothoid paths	47
2.12	Path with clothoid arc geometry	49
2.13	Euler interpolation	55

2.14 Comparison of a Dubins path with a Pythagorean hodograph path. The Dubins path $(- \cdot -)$ is the shortest path but it lacks the curvature continuity. The PH path (——) has continuity but is longer for the same curvature bound   59
2.15 Evolution of a PH path from the tangent continuity into curvature continuity   62

3.1 Three-dimensional Dubins manoeuvre conditions   66
3.2 Three-dimensional Dubins manoeuvre of a UAV   67
3.3 Spatial PH path with tube   78

4.1 Obstacle avoidance in 2D   83
4.2 Single-obstacle Dubins   86
4.3 Single-obstacle Dubins line intersection   87
4.4 Single-obstacle Dubins line solution   88
4.5 Single-obstacle Dubins line solution limit condition for the same rotation direction   89
4.6 Single-obstacle Dubins line solution limit condition for opposed rotation direction   90
4.7 Single-obstacle Dubins intersecting line solution with the same rotation direction   90
4.8 Fail condition for same rotation direction   93
4.9 Fail condition for opposed rotation direction   94
4.10 Solution for scaling start arc segment   94
4.11 Arc intersection   95
4.12 Arc intersection tangent vectors   96
4.13 Arc intersection sufficient conditions   97
4.14 Arc scaling for maximum-curvature intersection   98
4.15 Arc scaling for minimum-curvature intersection   98
4.16 Solution set for arc intersection   99
4.17 Multiple obstacle environment   99
4.18 Multiple obstacle intersection trajectories   100
4.19 Clockwise Dubins trajectory   101
4.20 Multiple obstacle multiple trajectories for clockwise trajectory   101
4.21 Multiple obstacle concave trajectory for anticlockwise trajectory   102
4.22 Multiple obstacle complete trajectory set   102
4.23 Complete trajectory set for 15 obstacles   103
4.24 Threat handling by intermediate pose   105

4.25 UAVs in a static cluttered environment 106
4.26 Dubins flyable paths of two UAVs in a cluttered
environment 107
4.27 Re-planning the Dubins path of UAV2 by curvature
adjustment 108
4.28 Re-planning the Dubins path of UAV2 using
an intermediate waypoint 108
4.29 Five UAVs each with four waypoints in cluttered space 109
4.30 Initial paths (only tangent continuity) – PH 2D
in cluttered space 110
4.31 Flyable paths – PH 2D in cluttered space 111
4.32 Feasible (safe and flyable) paths – PH 2D in cluttered space 112
4.33 Paths of equal lengths – PH 2D in cluttered space 113
4.34 Obstacle avoidance in 3D 113
4.35 Path planning with obstacle avoidance in 3D 114
4.36 Obstacle avoidance in 3D (2D projection) 115

5.1 Guidance geometry 121
5.2 Carrot guidance 125
5.3 Linear guidance trajectory 127
5.4 Nonlinear guidance trajectory 131
5.5 Basic geometry for collision of UAV and aircraft 132
5.6 Basic geometry for collision avoidance of UAV and aircraft 133
5.7 Sightline geometry for single UAV and aircraft 134
5.8 Sightline miss geometry for single UAV and aircraft 135
5.9 Geometry for single UAV and aircraft 136
5.10 Chord intersection for sector definition 136
5.11 Velocity geometry 139
5.12 Relative position geometry 140
5.13 Collision condition for single UAV and two aircraft 142
5.14 Sightline miss geometry for single UAV and two aircraft 143
5.15 Chord intersection for multiple conflict resolution 144
5.16 Avoidance algorithm trajectories for multiple aircraft 145

6.1 UAV safety and communication range spheres 148
6.2 Multiple UAVs scenario: $r_i(t)$ are flight paths, (x, y, z, θ, ψ)
are poses. Suffix i represents the ith UAV or path. Shaded
regions are obstacles/threats 149
6.3 Path planner for flyable paths 151
6.4 Safety constraints for collision avoidance 153

6.5   UAV shortest flyable paths – Dubins 2D   158
6.6   Paths of equal length – Dubins 2D   159
6.7   Separation distance for paths of first four
       combinations – Dubins 2D   160
6.8   Separation distance for paths of second four
       combinations – Dubins 2D   161
6.9   Separation distance for paths of last two combinations –
       Dubins 2D   161
6.10   Initial flyable paths of UAVs – clothoid 2D. All UAVs have
       different path lengths and the paths are intersecting
       with one another   162
6.11   Final paths of UAVs (paths of equal lengths) – clothoid 2D   163
6.12   PH paths of equal lengths   164
6.13   PH Paths of UAVs, equal lengths elevated at constant altitude   164
6.14   Flyable paths of UAVs – Dubins 3D   166
6.15   Paths of equal lengths – Dubins 3D   167
6.16   Paths of equal length (UAV1 and UAV2) – Dubins 3D   167
6.17   Paths of equal length (UAV1 and UAV3) – Dubins 3D   168
6.18   Paths of equal length (UAV2 and UAV3) – Dubins 3D   168
6.19   Flight path intersections. The intersections are given in each
       plane for each UAV. The intersections are calculated
       numerically. This is to avoid the possibility of complex points
       during the intersection between lines and circles. The following
       safety conditions are tested individually on all planes:
       (i) minimum separation distance, and (ii) non-intersection
       at equal lengths   169
6.20   Curvature and torsion variations with respect to path length,
       tangent-continuous path – UAV1. The path does not meet
       the maximum-curvature bounds at the boundaries   170
6.21   Curvature and torsion variations with respect to path length,
       tangent-continuous path – UAV2. The path does not meet the
       maximum-curvature bounds at the boundaries   171
6.22   Initial paths (curvature-continuous and flyable) for UAV1
       (solid line) and UAV2 (dotted)   172
6.23   Curvature and torque variation of flyable path r_1
       of UAV1 – PH 3D   172
6.24   Curvature and torque variation of flyable path r_2
       of UAV2 – PH 3D   173

6.25 Final, feasible paths (solid) – UAV1 and UAV2
      (dashed line is the initial path of UAV2)   173
6.26 Curvature variation: feasible path of UAV2 – PH 3D
      (length is equal to that of UAV1)   174

A.1   Frenet–Serret frame $\{t, n, b\}$, in which t is the unit tangent,
      n is the unit normal and b is the unit binormal. On the diagram,
      $r(q)$ is the path, P is the position vector of a point on the path,
      $\{e_x, e_y, e_z\}$ are the unit vectors and $h(q)$ is the path length   177
A.2   Frenet–Serret frame on a 3D curve   180

List of Tables

2.1 Calculation of tangent exit and entry points for external tangent   35
2.2 Calculation of tangent exit and entry points for internal tangent   39
2.3 Initial variables   54

Nomenclature

v	UAV velocity vector
v	UAV speed (m s^{-1})
θ	Horizontal angle in inertial axes (rad)
ψ	Vertical angle in inertial axes (rad)
ϕ	Dubins arc angle (rad)
x, y, z	Inertial position coordinates (m)
κ	Curvature (m^{-1})
τ	Torsion (m^{-1})
ρ	Radius of curvature (m)
$r(q)$	Path, with path variable q
h	Path length (m)
e	Basis axes vector set
t, n, b	Tangent, normal and binormal vectors, which define a basis axes set in e
$\mathbf{R}$	Rotation matrix
$\dot{\zeta}_s$	Sightline rate (rad s^{-1})

1

Introduction

Autonomy is replacing the human operator in many applications. Examples involve military systems where there is some element of danger to the human operator, civilian systems when handling hazardous materials, as well as monotonous operations such as surveillance, reconnaissance and *dull, dirty and dangerous* missions, such as operations in chemical and biological environments (Blyenburgh 1999; NMAB and ASEB 2000) and in environmental monitoring (Roberts *et al.* 2008). The replacement of the human operator in such systems necessitates the development of autonomous systems techniques. Such autonomous systems operate in many environments, such as in the air, in and under water, in space or on the land.

In this book, unmanned aerial vehicles (UAVs) are studied, operating as a group. The large range of potential applications of UAVs in military and civilian sectors have generated a lot of academic as well as commercial research (OSD 2005; Wilson 2007). Inspired by examples in nature, such as flocks of birds, shoals of fish, swarms of bees, and colonies of ants, cooperative control (Rabbath *et al.* 2004; Uny Cao *et al.* 1997) has become one of the active research areas in autonomous systems. Employing a group of UAVs rather than a single UAV can result in cost-effective and fault-tolerant systems. Advances in avionics, navigation based on GPS (Global Positioning System), flight control techniques and low-cost electronics have further fuelled the use of UAVs in commercial and military applications. Future UAVs will be more autonomous than the remotely piloted reconnaissance platforms in use today.

Cooperative Path Planning of Unmanned Aerial Vehicles
Antonios Tsourdos, Brian White and Madhavan Shanmugavel
© 2011 John Wiley & Sons, Ltd

One of the open issues in the development of autonomous systems is that of path planning. A path planning algorithm produces one or more safe flyable paths for the UAVs. The path has to be of a specified (usually minimal) length, and, as the UAV has limited range, the time spent surveying specific areas should be minimised. In addition, when surveying an area or a location, it is beneficial to be able to approach from specific directions in order to minimise obscuration as well as to aid identification. Hence the path length and direction will always be major factors in any path planning algorithm.

The UAVs should be capable of following any resulting path. This implies that the trajectory must comply with the speed and manoeuvre constraints of the UAVs. The path planning algorithms must also allow for the deployment of several UAVs in a coordinated manner, which will involve collision avoidance and simultaneous arrival at one or more locations. Finally, path planning algorithms are required to be coded in software that runs on a processor carried on-board the UAVs. Thus, they must be computationally efficient, take up a small amount of memory and operate in real time, enabling the UAV to re-plan its trajectory if needed, with no significant delay.

1.1 Path Planning Formulation

The primary aim of path planning is to provide structured mobility, that is, to facilitate moving or flying multiple UAVs from one location to another. There may be several locations to visit before reaching the final destination, and hence several consecutive paths may be required. Generally, there will be several predefined points of interest (POIs) on a known or partially known map/area. The UAV will have a specific attitude, which is combined with its location to give the UAV pose $P(x, y, z, \theta, \psi)$, where (x, y, z) is the UAV location or waypoint and (θ, ψ) are the horizontal and vertical angles, respectively. Consider a UAV moving from one pose, P_s, to another, P_f, where P_s and P_f are labelled the start and finish poses, respectively. Path planning involves producing one or more flight paths $r(q)$ connecting P_s and P_f. Mathematically, this can be represented as

$$P_s \xrightarrow{r(q)} P_f, \tag{1.1}$$

where $r(q)$ is the resulting path, and q is defined as a path parameter. This parameter can be a length variable $(0 \leq q \leq s)$ for a straight-line path or an angle variable $(0 \leq q \leq \theta)$ for a curved path. The choice of path variable depends on the path formulation.

Equation (1.1) is in a very simple form. For a single UAV flying from a location with start pose $P_s(x_s, y_s, z_s, \theta_s, \psi_s)$ to a location with finish pose $P_f(x_f, y_f, z_f, \theta_f, \psi_f)$, equation (1.1) can be written in the form

$$P_s(x_s, y_s, z_s, \theta_s, \psi_s) \xrightarrow{r(q)} P_f(x_f, y_f, z_f, \theta_f, \psi_f). \tag{1.2}$$

Extending equation (1.2) for N UAVs, where each pair of poses are connected by paths $r_i(q)$, gives

$$P_{si}(x_{si}, y_{si}, z_{si}, \theta_{si}, \psi_{si}) \xrightarrow{r_i(q)} P_{fi}(x_{fi}, y_{fi}, z_{fi}, \theta_{fi}, \psi_{fi}), \qquad i = 1, \ldots, N. \tag{1.3}$$

Equation (1.3) connects a pair of points by a path. This problem is well known as route planning in the fields of operations research, communications, computational geometry and graphics, where a route is generated between one or more nodes of a network. However, applying the route planning concept from these fields to flying vehicles becomes challenging. The route is usually defined by a set of waypoints joined by straight-line segments, which connect the start and finish waypoints, and hence may not be flyable because the UAV cannot turn instantaneously through each waypoint. For a flyable path, each segment must have a common tangent to produce a continuous path. Hence it is important to specify the orientations at each waypoint that each segment must match. This implies that some segments must be curved rather than straight in order that each end of the segment meets the common tangent condition. Also, some missions – such as target acquisition, search and track, and disaster area surveying – require the sensors to be pointed in specific directions for effective detection and identification.

1.2 Path Planning Constraints

Producing a path between the start and finish poses is straightforward in the absence of any constraints. In practice, there are various constraints involved in path planning, most being UAV-specific and the remainder arising from obstacles in the environment. Sometimes the violation of constraints, for example, communication failure, may lead to the complete loss of the UAV in extreme circumstances and to the loss of the UAV as a sensor platform in less extreme cases. Hence, safety of the UAVs is important throughout the mission. Other constraints such as the minimum turn radius will also dictate which paths are flyable and which are not.

Hence, the two most important constraints for path planning of a UAV are that the path must be flyable and safe. Flyable paths meet kinematic or motion constraints and dictate the manoeuvrability of the UAVs. The safety of the UAV is achieved by avoiding obstacles, either fixed or moving, that intersect the path. The obstacles may be UAVs, aircraft and buildings, which are common in airspace and in urban environments. Other constraints, such as maintaining communication range, as well as minimising time and/or path length, can be added into the system where necessary. We use the symbol $\coprod$ to represent the constraints. Hence the constrained path planning can be written in the form

$$P_{si}(x_{si}, y_{si}, z_{si}, \theta_{si}, \psi_{si}) \xrightarrow{\;\coprod r_i(q)\;} P_{fi}(x_{fi}, y_{fi}, z_{fi}, \theta_{fi}, \psi_{fi}). \qquad (1.4)$$

From equation (1.4), we can make an analogy that the path planner acts like a black box, which produces the flyable path from given inputs as shown in Figure 1.1. The inputs are the poses, with additional constraints, uncertainties and measurements. The feedback loop senses the measured states of the UAV and also feeds back the success in terms of meeting the constraints to the path planner.

1.2.1 Flyable Paths: Capturing Kinematics

As the properties of the path influence the motion of the UAV and vice versa, it is important to understand the characteristics of the path. The UAV needs a path in order to move from one location to another. However, the path has to meet the dynamic constraints of the UAV. In order to understand

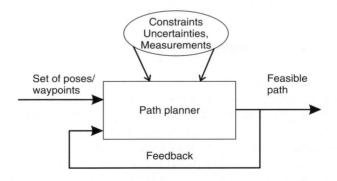

Figure 1.1 A block diagram approach to path planning

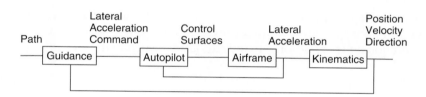

Figure 1.2   Autopilot and guidance control loops

the dynamic constraints of the UAV and hence the kinematic constraints of the path, consider the UAV as a control system. Two control loops are present in the UAV system, as shown in Figure 1.2. The inner loop is known as the autopilot and the outer loop is known as the guidance system. The guidance system provides lateral acceleration commands to keep the UAV following the path, whereas the autopilot controls the UAV elevator, ailerons and rudder to achieve the required lateral acceleration.

The UAV dynamics include the aerodynamics, which produces forces and torques on the airframe. From Newton's third law of motion, the forces and torques produce lateral, longitudinal and rotational accelerations. The accelerations are usually expressed in UAV body axes and these provide the link to the kinematics. The kinematics is produced by integrating the UAV lateral and rotational acceleration vectors to obtain the UAV velocity vector. The attitude angles and the current UAV position in the inertial frame give the UAV's translational and rotational velocity. For example, the two-dimensional (2D) path planner uses the kinematic model in equation (1.5). The kinematics and hence the current state of the UAV in 2D are thus

$$\dot{x} = |v|\cos(\theta), \tag{1.5a}$$

$$\dot{y} = |v|\sin(\theta), \tag{1.5b}$$

where v is the UAV velocity and θ is the horizontal heading angle.

Whether a given path is flyable or not is determined by the curvature of the path. The path planner has to produce a path $r(q)$ that meets the dynamic turn rate constraint of the UAV, which is translated into the kinematic curvature constraint – in 3D, it is determined by both the curvature and torsion (Lipschutz 1969). By satisfying this constraint, the motion of the UAV stays within its maximum-curvature (acceleration) bounds. The curvature is proportional to the lateral acceleration. Thus the curvature at any point on the path must be less than the maximum-curvature constraint of the UAV.

A curve segment of zero curvature is a straight line and a curve segment of constant curvature is an arc of a circle. At a given speed v, the lateral acceleration a is proportional to the curvature κ such that

$$|a| = |v|^2 \kappa \propto \kappa, \qquad (1.6)$$

where a is the lateral acceleration vector, κ is the curvature and v is the velocity vector. Note that, for a constant speed, the acceleration vector a is normal to the velocity vector v. This ensures that the velocity vector rotates without changing its magnitude. This is illustrated in Figure 1.3, where the velocity vector is aligned with the tangent vector t and the acceleration vector is aligned with the normal vector n.

1.2.2 UAV Inertial Manoeuvre Coordinates

All manoeuvres can be defined by reference to a set of axes. These can be fixed in inertial space or defined relative to a set of axes that are attached to the path defined by a curve known as the Frenet–Serret (FS) framework, which will be dealt with later in the book. A set of axes may be defined that comprises a tangent vector t to the curve, together with a normal to this tangent n. Together with a binormal vector b, these make up a right-handed triple, which can be used to define the manoeuvres of the UAV. Two-dimensional path planning confines the manoeuvre of the UAV to a plane. If the plane is horizontal, this is equivalent to flying at a constant altitude. The plane need not be horizontal, but can be inclined at an angle to facilitate a change in altitude if required. However, in reality, UAVs fly in three-dimensional (3D) space, and so path planning needs to be able to produce flyable paths in three dimensions. In 3D, flyable paths need to accommodate both curvature and torsion in their design. Curvature defines the turn radius of the path in 2D, which is defined as a rotation about an axis normal to the manoeuvre.

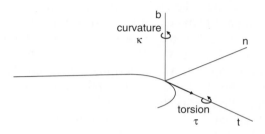

Figure 1.3　Curvature and torsion

Torsion is defined as a rotation about an axes that coincides with the path tangent vector. This is shown in Figure 1.3. Note that positive curvature and positive torsion are defined as being clockwise along the appropriate axis when looking along the axis from the origin.

For the UAV, in terms of body axes, curvature is equivalent to a yaw rate turn and torsion is equivalent to a roll rate turn. However, in practice, the UAV will perform a turn by first rolling to a fixed bank angle and then using the elevator to produce a manoeuvre normal to the wings. This is known as a bank-to-turn manoeuvre.

1.2.3 Generation of Safe Paths for Path Planning

The second important constraint of path planning is safety. The planned path should be flyable and also safe. Safety is measured by the ability of the UAV to avoid fixed and moving obstacles and other UAVs. As the UAV follows the path defined by the path planner, the safety constraint has to be defined with respect to the properties of the path. The path must maintain collision avoidance with other friendly UAVs and also must be flexible enough to avoid obstacles. The safety of the UAV is represented by $\coprod_{\text{safe}}$, and can be used in equation (1.4). Hence, taking into account the curvature constraint $\coprod_{\kappa}$ and safety, the equation of path planning is now modified to

$$P_{si}(x_{si}, y_{si}, \theta_{si}, \psi_{si}) \xrightarrow{\coprod_{\text{safe}}, \coprod_{\kappa}, r_i(q)} P_{fi}(x_{fi}, y_{fi}, \theta_{fi}, \psi_{fi}). \qquad (1.7)$$

There may be other constraints, such as maintaining communication in a complex urban environment, as well as time, task completion and resource management, depending on the mission objectives. The subscript can be changed for different constraints. For example, imposing a communication constraint on UAV separation distance would be represented by $\coprod_{\text{comm}}$.

1.3 Cooperative Path Planning and Mission Planning

The previous section defines path planning for individual UAVs. Cooperation involves using the path planning algorithms to produce a coordinated mission. One such mission is for a group of UAVs to set off from a base and reach a rendezvous position at the same time. The UAVs will perform tasks on the way, such as area searches and object detection.

The payload, sensor suite and duration of operation, together with the cost and size of the UAV, are important in their use and deployment. The limited sensor range, period of operation and payload limitations usually result in the need to use more than one UAV. The mission can be accomplished faster if a single UAV is replaced with a group of UAVs. Such operations do not simply involve operating together in an environment. The challenging task is cooperation among the UAVs, and coordination of their activities. Therefore, a group of UAVs of the same (homogeneous) or different (heterogeneous) capabilities can act together as a single entity to achieve the mission. Also, a system of cooperative UAVs is fault-tolerant, and the performance is better than that of a single UAV.

For the case of autonomous UAV systems, cooperation between UAVs means shared utilization of resources, sharing of information, allocation of tasks, coordination of actions and operations, coping with disturbances and dealing with conflict resolution. The level of cooperation can be set by a central decision-maker in a ground control station, for example, or it may be left to individual UAVs in a distributed manner. Also, it is important to note that cooperation is a complex problem. Consider collision avoidance between two cooperating UAVs. Though they mutually manoeuvre to avoid collision, the resulting modified path plans may lead to collision with other UAVs in the group. Therefore, cooperation has to be planned with respect to the other members of the UAV group, as well as obstacles (both fixed and moving) in the environment and uncertainties. It has been argued (Ren *et al.* 2004) that information sharing is central to cooperation and coordination, where information sharing is represented by a coordination variable. However, a minimum amount of information for cooperation has to be defined for the system. This, in turn, affects the cost of communication and computation. Cooperative behaviour is also studied in robotics research (Fong *et al.* 2003; Howard *et al.* 2006; Uny Cao *et al.* 1997). Other related research areas are 'multi-agent control', 'distributed networks', 'consensus algorithms', 'cooperative control', 'network control' and 'swarm intelligence' (Qu 2008; Shamma 2007). All these research areas emphasise that the sharing of information is the important factor in cooperative system. A variety of other applications, such as task allocation (Beard *et al.* 2002), flight formation (Fowler and D'Andrea 2003), surveillance, suppression of enemy air defence (SEAD) and radar jamming, have been studied for the cooperative system in recent research. An overview of recent research on cooperative systems can be seen in Murray (2007).

Cooperation can be achieved by central coordination or distributed decision-making. Can cooperation be achieved simply by solving a set of equations or by taking decisions based on testing conditions and/or constraints? This problem is still an open research area for autonomous systems.

For example, collision avoidance between two flight paths $r_1(q)$ and $r_2(q)$ can be represented by

$$r_1(q) \cap r_2(q) = \varnothing. \tag{1.8}$$

This equation captures the condition required to be satisfied for collision avoidance. A cooperative path planning of multiple UAVs for collision avoidance can be formulated as producing flight paths $\{r_1(q), r_2(q), \ldots, r_n(q)\}$ subject to the following constraints:

$$\sqcup = \begin{cases} \sqcup_\kappa, & \kappa \leq \kappa_{\max}, \\ \sqcup_\tau, & \tau \leq \tau_{\max}, \\ \sqcup_{\text{safe}}, & r_i(q) \cap r_j(q) = \varnothing, \qquad i \neq j. \end{cases} \tag{1.9}$$

Such problems can be solved by an optimisation algorithm, but this involves heavy computation. Owing to the complexity of the problem, in many cases the problem is divided into different phases/stages. The optimal or suboptimal solution will be a sequential solutions to the subproblems.

As cooperative behaviour arises when two or more UAVs are involved, a functional architecture is necessary to coordinate the communication and

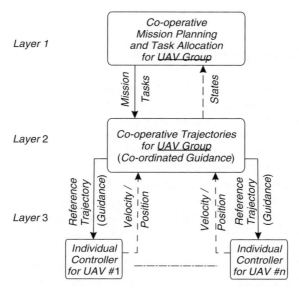

Figure 1.4   Hierarchy of mission planning. Reprinted with permission of Elsevier, and ASME

control of each UAV. One such architecture is shown in Figure 1.4. The cooperative behaviour is defined by the central mission planner or can be in-built into each UAV, depending on the autonomy architecture. As shown in the figure, the path planning activity is a subsystem of the mission planner. The figure shows three subsystems or layers. However, the number of subsystems and their functions may vary for different applications and mission objectives. The top layer holds and keeps track of the objective(s) of the mission. Based on these objectives, this layer allocates resources and tasks to the UAVs and also acts as decision-maker. The intermediate (second) layer produces trajectories (paths) for the UAVs. In this layer, the path planning and their associated algorithms such as collision avoidance to produce feasible trajectories (paths) are located. The lower (third) level produces guidance and control actions, which ensure that the UAVs fly on the reference trajectories produced by the second level. This book focuses on the second layer, where the path planner produces flight trajectories (paths) to fulfil the mission objectives. The mission objective considered here is the simultaneous arrival at a specified location by a group of UAVs. Note that one of the main requirements for successful cooperation and coordination is that the UAVs have to be in communication most of the time during the mission. This is dealt with using the communication constraint $\amalg_{comm}$ mentioned in section 1.2.3.

1.4 Path Planning – An Overview

Path planning is an integral part of an autonomous system, which is responsible for moving from one point to another. An autonomous system may be operating either on land, in the air, or on/in water. The paths are designed by various techniques based on whether the system has to traverse an area that is known, unknown, or partially known. The physical limitations, operating environment and communication requirements make the path planning more complex. It is essential to have an on-board processor to design and execute paths and trajectories. Recent work in cooperative application of unmanned systems has employed multiple autonomous vehicles to enhance the effectiveness of operation success. Such applications further increase the complexity of the path planning.

Research into path planning is widely documented in ground robotics and manipulator systems. Most of the path planning techniques that are currently in use are borrowed from the ground robotics field. Originally, research on ground robotics focused on indoor robots. The lessons learned from them were applied to outdoor autonomous vehicles, and to aerial robotics, with improvements and modification. In all these applications, path planning plays an integral part and an important role. This is described in various references, for example, Bender (1991) and Shladover (1991) in ground robotics, Chan and Foddy (1985), Hebert *et al.* (2001), Vian and More (1989) and Zabarankin *et al.* (2002) on aerial vehicles, Yuh (1995, 2000), Smith *et al.* (2001) and Oliveira *et al.* (1998) in underwater vehicles, and Tompkins *et al.* (2004) and Gennery (2004) in space. Therefore, it is appropriate to review the techniques originally developed for ground robotics.

The objectives and approaches of path planning differ depending on the application domain: surveillance, search and track, rescue missions and disaster monitoring. Currently, there are a multitude of solution approaches available in the research literature, and each approach has its own merits. Moreover, new approaches are required as the complexity of the problem increases. However, the majority of solution approaches can be represented by a simplified diagram as shown in Figure 1.5. Using this approach, the inputs to the path planner are the waypoints, the obstacle positions and size, and the associated uncertainties. Optimisation techniques are applied to these data to produce routes. The resulting routes are usually polygonal paths, which do not have inherent curvature constraints built into the solution. In most cases the path planner produces routes that are applicable to mobile robots as they can virtually stop and turn. But in the case of UAVs, these

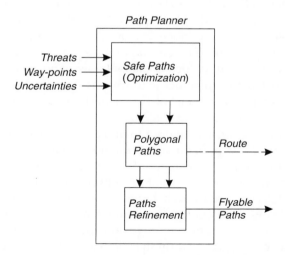

Figure 1.5 Existing approach to path planning. Reprinted with permission of Elsevier

routes may not be flyable, driveable or manoeuvrable. A flyable path meets the kinematic constraints and the imposed dynamics of the robot. Therefore, attention has turned towards the development of paths that can be flyable or manoeuvrable. Therefore, these routes have to be further refined to produce flyable paths.

In the literature, a wide variety of approaches are used to produce routes using a route planner. The route planner produces a series of nodes that are connected by straight lines, in turn connected to the start and finish points. This is also called a 'global path planner', and produces one or more routes for the given map of known locations. The locations may be places to visit or places to avoid. The route is produced by isolating the obstacles, no-fly zones and threats. The road map, and optimization methods that produce routes using straight-line segments are route planners. Though the routes cannot be followed by the UAVs directly, they are important, as they can be produced faster than producing a flyable path in an environment containing fixed obstacles. Hence, it is common for the path planner first to produce a route and then later to refine the route for a flyable path. However, this is not the only way to produce the path.

Path planning approaches can be divided into several categories, based on different criteria. However, the nature of the applications, environment, medium of operation and path constraints have produced a variety of algorithms and techniques. The predominant methods used in ground robotics and documented in Latombe (1991) are: (i) the road map method, (ii) the cell decomposition method, and (iii) the potential field method. These methods generate routes for a robot to move from a start point to a finish point. Also, these methods rely on specific definition of the environment. The environment is defined by a map, which contains known and unknown obstacles. Finding a path between two points on the map is simplified by (i) discretizing the map into small areas or cells or (ii) converting the map into a continuous field. Hence, path planning can classified as discrete or continuous. A suitable search algorithm is then used to find a path connecting the start and finish points on this simplified map. The road map and cell decomposition methods transform the environment into a discrete map, while the potential field method transforms the map into a continuous function. The textbook by Latombe (1991) is a thorough review of these methods. Here, only a descriptive version is presented for completeness. These methods, which produce a path for a given map or environment, are also called global path planners. These methods transform the given environment into a searchable database. We consider each in turn.

1.5 The Road Map Method

A road map itself is not a path planner. It is a two-dimensional network of straight lines connecting the initial and final points without intersecting the obstacles defined in the map. In other words, it is a representation of the 2D environment by a non-directional graph. The road map methods work in configuration space, where the robot is treated as a point while the workspace is modified to accommodate the physical size of the vehicle. For a given start point P_s and finish point P_f, all possible connecting paths are generated by avoiding obstacles. A typical metric such as A* search algorithm (LaValle 2006) is used for generating the shortest route between destinations. The resulting route consists of a series of waypoints connecting the start and finish locations.

In general, the algorithms work in the following way (see Figure 1.6). The map is defined by a space named the total space, C_{space}. This is then split into the free space (obstacle-free space) C_{free} and the obstacle space C_{obst}. Next, a network of graph connectivity Q_{free} is generated by selecting a set of points that can be connected by straight lines such that the resulting discretization produces a set of polygons that surround each obstacle in C_{space}. The resulting connectivity graph is used to produce a network of all

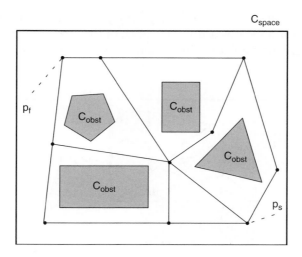

Figure 1.6 The road map method

possible collision-free routes. Search algorithms such as A* are then used to find one or more routes based on some metric between the start and finish poses or locations. Two such methods used in the literature employ the connectivity graph to produce a route. These are (i) the visibility graph, and (ii) the Voronoi diagram.

1.5.1 Visibility Graphs

As the name suggests, the visibility graph produces a line-of-sight route through an environment. This is one of the earliest methods used for path planning. The route is formed by a connectivity graph network of a non-directed graph of straight lines. For this method, only polygonal obstacles are considered. For a graph $G = (V, E)$, the vertices V are the vertices of the obstacles, while the edges E connect all the vertices with straight lines, provided the lines do not intersect the obstacles. Hence only vertices that are visible in the sense that each vertex can be seen from the other are included. A schematic sketch of a visibility graph is shown in Figure 1.7. A route is then found using a graph search algorithm, which connects the start and finish locations, treating them as vertices. Hence only visible vertices are connected to the start and finish vertices.

1.5.2 Voronoi Diagrams

A Voronoi diagram is a connectivity graph generated by forming polygons or 'fences' around the obstacles. Each edge of the fence polygons is defined by first constructing a set of lines connecting the centres of the obstacles.

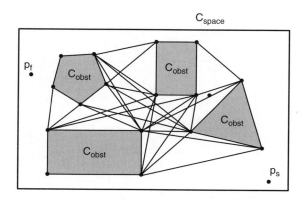

Figure 1.7   Visibility graph

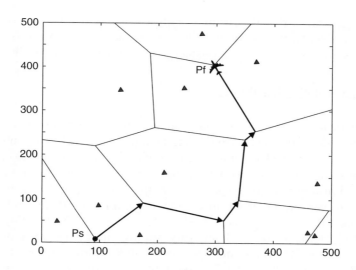

Figure 1.8   Voronoi diagram: polygonal fences around obstacles

The set of polygonal fences is then constructed by drawing a set of lines perpendicular to the lines connecting the obstacles. These are then adjusted to meet at a minimum set of vertices. An example of a Voronoi diagram is shown in Figure 1.8.

The resulting polygonal fences can be considered as a connected graph. A search algorithm such as A* can be used to find a route connecting an initial and final vertex in the graph. The Voronoi diagram is used in Chandler *et al.* (2000) to produce a route for a UAV flying in an environment of static radar sites whose locations are known *a priori*. The route is refined by adding fillets at the vertices. Simultaneous arrival of multiple UAVs is coordinated by a high-level manager based on the sensitivity function (cost versus time of arrival) broadcast by each UAV. A similar approach is adopted in Bortoff (2000), where an analogy of a chain connected by sequences of spring–mass–damper systems to the UAV path is used. The ends of the chain are located at the initial and final configurations. The obstacles induce a repulsive force, which causes the masses in the chain to move away from the obstacles. However, this method involves complexity in solving ordinary differential equations (ODEs) with curvature constraints. Also, the accumulation of only a few masses around the obstacle location will lead to coarse path resolution, which is undesirable. In McLain and Beard (2000), the above approach is extended by replacing the spring–damper system with rigid links between masses to eliminate sharp corners. However, this method does not guarantee that the resultant path is flyable by a UAV. Later (Judd and McLain 2001), the Voronoi path is interpolated with a series of cubic splines, assigning a cost to each obstacle location (Chan and Foddy 1985).

1.6 Probabilistic Methods

Probabilistic methods work by a random selection of neighbourhood points that meet some metric such as the shortest length resulting in the probabilistic random road map (PRR) method. This method samples the given space for probable solutions in the form of a network of graphs. It normally uses uniform sampling of a given space. The path planning problem is treated as a search problem in the partitioned cells (Eagle and Yee 1990). Probabilistic road maps (PRMs) (Kavraki *et al.* 1996; Pettersson and Doherty 2004) connect the starting point to the goal point by adding successive trajectories to a pre-computed route. In another approach called rapidly exploring random trees (RRTs) (Cheng *et al.* 2001; LaValle 1998), a tree of trajectory segments is extended from the start point to the goal point. Every successive trajectory in the tree is selected randomly by connecting to a closest point in the existing tree. In Amin *et al.* (2006), the path planning is achieved by the RRT and is further enhanced by using the Dijkstra search algorithm, which finds the route for the UAV flying among known static obstacles represented as quadtrees. Probabilistic methods are applied to path planning by considering the positional uncertainty of threat regions in Jun and D'Andrea (2003). The final path is refined with circular arcs at the points of line joining.

1.7 Potential Field

The potential field method was first proposed by Khatib (1985). In this method, the environment is represented as an artificial potential field. The destination point is assigned an attractive potential, while the obstacles are assigned repulsive potentials. The idea is that a robot moving in the field will be attracted towards the destination, while being repelled by the obstacles. In contrast to the road map and cell decomposition methods, the path resulting from this method follows the line of maximum potential of a continuous field function. However, this method has some drawbacks, the main one being that the vehicle may get trapped in a local maximum, for example, when encountering a C-shaped obstacle (Borenstein and Koren 1991; Koren and Borenstein 1991). However, improved versions of this method have been developed to eliminate the local maxima and to reduce the computational complexity. A variation of the technique using multiple temporary attraction points together with genetic algorithms (GAs) are used in Arámbula and Padilla (2004). The potential field algorithm (Eun and Bang 2006) solves the path planning by generating an attractive field towards the goal point and a repulsive field at the obstacles. In another approach (Polymenakos *et al.* 1998; Tsitsiklis 1995), a Dijkstra-like method (Dijkstra 1959) is suggested for solving a continuous-space shortest-path problem in a 2D plane by optimization. An analytical and discrete optimization approach has been used (Zabarankin *et al.* 2002) for optimal risk path generation in 2D space with constant radar cross-section, arbitrary number of sensors and a constraint on path length.

1.8 Cell Decomposition

In the cell decomposition method, the environment is divided into non-overlapping cells. Possible routes are then generated that pass through adjacent free cells. Free cells are ones that are not occupied by obstacles. The obstacles are isolated by finding the connectivity between the free cells. Thus a discrete version of the environment is produced. A search algorithm is used to connect adjacent free cells. Figure 1.9 shows the schematic of the process. The shaded (cross-hatched) cells are eliminated because they are occupied by obstacles (grey). A connectivity between the start and finish points is found by connecting the free cells by a series of straight lines.

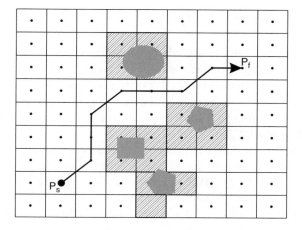

Figure 1.9   Cell decomposition

The figure shows a simple division of the environment, which is called the exact method of cell decomposition. An approximate method is proposed to reduce the memory consumption in Kambhampari and Davis (1986). This is also called the quadtree or octree representation. Further improvements of this method can be found in Jung and Gupta (1996).

1.9 Optimal Control

Optimal control is the most natural way to solve problems involving objective functions, constraints and boundary conditions. However, the dimension and complexity of optimal control problems cause a heavy burden on computational time in the solution. Also, the nature of the problem may require either suboptimal or feasible solutions rather than the optimal solution. It is quite difficult to solve path planning problems in a cluttered environment using optimal control theory (Duleba and Sasiadek 2003; Sasiadek and Duleba 2000). General planar motion is a combination of straight-line and circular trajectories, which are proved for a mobile robot using Pontryagin's maximum principle in Balkcom and Mason (2000). Optimization techniques such as probabilistic methods, mixed integer linear programming and genetic programming have been applied to path planning of UAVs. These techniques produce paths by optimizing certain cost functions. The cost functions differ based on the applications, such as minimum time of arrival, optimizing fuel consumption and coordinated motion. They are mostly search algorithms.

1.10 Optimization Techniques

The challenges and complexity of path planning with optimization techniques and uncertainties are discussed in Patcher and Chandler (1998). In Rabbath et al. (2004) an overview of coordinated control of UAVs and their complexities is presented. Branch and bound optimization is used for path planning and is compared with mixed integer linear programming (MILP) in Eele and Richards (2009).

The use of MILP for path planning applications can be found in Ademoye et al. (2006), Schouwenaars et al. (2001, 2006), Richards et al. (2001) and Richards and How (2002). MILP is an application of the operations research method, called linear programming with integer or binary constraints. These constraints are used for logical decisions, such as turn left and move up. This method produces a safe route for UAVs. But the route has to be smoothed further to make it flyable. Also, the optimization methods are associated with long computational time. Accomplishing the mission objectives with the physical and functional limitations of UAVs further increases the complexity of the solution to the path planning problem (Mitcheel and Sastry 2003; Yang and Kapila 2002).

Evolutionary algorithms are used in Mittal and Deb (2007), where the path length, risk of collision and limits on maximum and minimum heights are optimized in path planning of a UAV. The discrete points generated by the planner are connected by B-spline curves for producing flyable paths. Also the evolutionary optimization principle in path planning can be seen in Dong and Vagners (2004), Zheng *et al.* (2005) and Nikolos *et al.* (2003). A genetic algorithm is used in Wu *et al.* (2007) for path planning of autonomous underwater vehicles for surveying a given area.

1.11 Trajectories for Path Planning

The optimization methods, randomized search approaches and Voronoi diagram approach use exhaustive search and computational methods that result in route planning. The route planning does not consider the kinematic constraints of the path. Also, the reactive behaviour of the UAV needs a flyable path at any point of its flight. In such a situation, route planning would be a handicap. For this reason, it appeared reasonable to attempt to use the curves directly in path planning. In this manner, planar and spatial Dubins paths (Dubins 1957; Shanmugavel *et al.* 2006b, 2007c), Pythagorean hodographs (Shanmugavel *et al.* 2006a, 2007b) and 2D clothoids (Shanmugavel *et al.* 2007a) have been used to solve the problem of simultaneous arrival on target. Though flyable paths are essential for manoeuvring, straight-line trajectories are used in other applications, such as task allocation for multiple robot problems in Zhang *et al.* (2008) and Shima *et al.* (2005). The Dubins path is used for airborne problems (Bicchi and Pallottino 2000; Massink and Francesco 2001; Robb *et al.* 2005; Shanmugavel *et al.* 2006b). The simultaneous arrival of UAVs using the Dubins path is studied in Shanmugavel *et al.* (2005, 2006b), where the path is produced using differential geometry. Parametric curves are used for path planning in Shanmugavel *et al.* (2006a) and Nikolos *et al.* (2003).

A comprehensive review of the types of paths used for path planning is presented in Segovia *et al.* (1991). Dubins (1957) showed in his work that the shortest path between two vectors in a plane that meets a minimum bound on turning radius is a composite path formed by line and circular arc segments. This paper received widespread attention by the research community and is extensively cited in ground robotics works (Bui *et al.* 1994; Latombe 1991; McGee *et al.* 2005; Shkel and Lumelsky 1996). This work has also motivated the derivation of the shortest path for a vehicle that can move forwards and backwards for parking a car. Reeds and Shepp (1990) developed the shortest path for a vehicle that can move both forwards and backwards. Clothoid arcs are used for the path planning of car-like vehicles using a composite path made of circular and clothoid arcs (Laumond 1986; Scheuer and Fraichard 1997). The Cornu spiral or clothoid curve has been used (Dai and Cochran 2009; Shanmugavel *et al.* 2009) for path planning of multiple UAVs. *B*-splines (Komoriya and Tanie 1989), quintic polynomials (Takahashi *et al.* 1989), polar splines (Nelson 1989), clothoids (Liscano and Green 1989), cubic spirals (Kanayama and Hartman 1989) and G^2-splines (Piazzi and Bianco 2000) have all been used for path planning of mobile robots. Robot path planning using the Voronoi diagram has been studied widely since the mid-1980s (Iyengar *et al.* 1986), and in the late 1990s the focus on coordinated path planning of multiple robots began. Much of the work done on path planning is carried out in ground robotics, but their approaches cannot be applied directly to unmanned aerial vehicles. The path of a UAV is limited by high-*G* turns and also by the fact that it has a threshold speed below which it cannot fly.

1.12 Outline of the Book

The book is divided into seven chapters. Chapter 1 introduces the importance of path planning for an autonomous system. Current research on path planning approaches is discussed in this chapter. As path planning is borrowed from the mobile robotics field, some focus on techniques used in this field are also discussed.

Chapter 2 deals with producing flyable paths in two dimensions. Three types of paths are considered. The chapter begins with the design of Dubins path, because this is the shortest path between two poses in 2D and also it is simple. The Dubins path is produced using the principles of Euclidean and differential geometries. Besides the equivalence of the results from the two approaches, it is shown that the differential geometric principles are advantageous in generalisation of the path. The lack of curvature continuity of the Dubins path motivates the use of other paths. In this respect, a single path (Pythagorean hodograph) and a composite path (clothoid and line segments) are considered. The circular arcs in 2D Dubins paths are approximated with clothoid segments to produce a smooth path. The last part deals with the Pythagorean hodograph (PH) curve known for its rational properties. A procedure is established to derive a PH path of curvature continuity.

Chapter 3 discusses three-dimensional path planning. It extends the principles used in Chapter 2. The Dubins path is produced in 3D using the principles of differential geometry. The clothoid path is not discussed, as the design is similar to that of the 3D Dubins path. The PH path is developed in 3D for use in path planning. The composite version of the Dubins path is generated by finding the common intersecting plane between the initial and final poses with an initial rotation at the start pose. The spatial PH path is developed with quaternion form and the curvature and torsion are met by increasing the tangent vectors at the initial and final poses.

Chapter 4 describes algorithms for detecting and avoiding threats or obstacles. Detection precedes avoidance. The chapter is divided into two major parts. The first part discusses obstacle avoidance in a static obstacle environment, while the second part discusses a strategy for avoiding dynamic obstacles. Obstacle avoidance in both 2D as well as 3D cases are considered. Two methods are used to avoid fixed obstacles: (i) decreasing the curvature, and (ii) generating intermediate waypoints. The methods are simulated for Dubins, clothoid and PH paths. For 3D path planning, a representative building area is considered for the simulation. The area is discretized to a searchable database. The Dijkstra algorithm is used to find the shortest route through the database. Following this, the waypoints are connected with flyable paths.

Chapter 5 presents a path-following algorithm for the Dubins path, which is required by the UAV to follow the feasible path. It develops two forms of guidance algorithm to follow Dubins and PH paths, using a simple linear guidance algorithm and a nonlinear guidance algorithm based on nonlinear dynamic inversion. The chapter then goes on to develop a guidance algorithm to avoid collision with mobile obstacles, which could be other UAVs or aircraft and other flying vehicles not under group control.

Chapter 6 discusses the problem undertaken for path planning, assumptions, problem formulation and solution description for the simultaneous arrival on target of a group of UAVs. An algorithm is used for the path planning. Handling the constraints involved by means of curvature variation is discussed. The flyable paths are tested for safety conditions. A solution is achieved by increasing the lengths over that of a reference path.

References

Ademoye, T. A., Davari, A. and Cao, W. 2006. Three dimensional obstacle avoidance maneuver planning using mixed integer linear programming. *Proc. Robotics and Applications 2006*, Honolulu, 14–16 August. ACTA Press.

Amin, J. N., Bošković, J. D. and Mehra, R. K. 2006. A fast and efficient approach to path planning for unmanned vehicles. *AIAA Guidance, Navigation, and Control Conf. and Exhibit*, Colorado, 21–24 August. AIAA 2006–6103.

Arámbula Cosío, F. and Padilla Castañeda, M. A. 2004. Autonomous robot navigation using adaptive potential fields. *Mathematical and Computer Modelling*, **40**(9–10), 1141–1156.

Balkcom, D. J. and Mason, M. T. 2000. Extremal trajectories for bounded velocity differential drive robots. *Proc. IEEE Int. Conf. on Robotics and Automation*, San Francisco, pp. 2479–2484.

Beard, R. W., McLain, T. W., Goodrich, M. A. and Anderson, E. P. 2002. Coordinated target assignment and intercept for unmanned air vehicles. *IEEE Transactions on Robotics and Automation*, **18**(6), 911–922.

Bender, J. G. 1991. An overview of systems studies of automated highway systems. *IEEE Transactions on Vehicular Technology*, **40**, 82–99.

Bicchi, A. and Pallottino, L. 2000. On optimal cooperative conflict resolution for air traffic management systems. *IEEE Transactions on Intelligent Transportation Systems*, **1**, 3578–3583.

Blyenburgh, van, P. 1999. UAVs: an overview. *Air & Space Europe*, **1**(5/6), 43–47.

Borenstein, J. and Koren, Y. 1991. The vector field histogram – fast obstacle avoidance for mobile robots. *IEEE Transactions on Robotics and Automation*, **7**, 278–288.

Bortoff, S. 2000. Path-planning for unmanned air vehicles. *Proc. American Control Conf.*, Chicago, pp. 364–368.

Bui, X. N., Boissonnat, J. D., Soueres, P. and Laumond, J. P. 1994. Shortest path synthesis for Dubins non-holonomic robots. *Proc. IEEE Int. Conf. on Robotics and Automation*, pp. 2–4.

Chan, Y. K. and Foddy, M. 1985. Real time optimal flight path generation by storage of massive data bases. *IEEE NEACON*, New York, pp. 516–521.

Chandler, P., Rasmussen, S. and Pachter, M. 2000. UAV cooperative path planning. *AIAA Guidance, Navigation, and Control Conf.*, Denver. AIAA-2000-4370.

Cheng, P., Shen, Z. and LaValle, S. M. 2001. RRT-based trajectory design for autonomous automobiles and spacecraft. *Archives of Control Sciences*, **11**, 167–194.

Dai, R. and Cochran, J. E. 2009. Path planning for multiple unmanned aerial vehicles by parameterized Cornu-spirals. *American Control Conf.*, pp. 2391–2396.

Dijkstra, E. W. 1959. A note on two problems in connexion with graphs. *Numerische Mathematik*, **1**, 269–271.

Dong, J. and Vagners, J. 2004. Parallel evolutionary algorithms for UAV path planning. *AIAA 1st Intelligent Systems Technical Conf.*, Chicago, 20–22 September.

Dubins, L. E. 1957. On curves of minimal length with a constraint on average curvature and with prescribed initial and terminal positions and tangent. *American Journal of Mathematics*, **79**, 497–516.

Duleba, I. and Sasiadek, J. 2003. Nonholonomic motion planning based on Newton algorithm with energy constraints. *IEEE Transactions on Control Systems Technology*, **11**, 355–363.

Eagle, J. and Yee, J. 1990. An optimal branch-and-bound procedure for the constrained path, moving target search problem. *Operations Research*, **28**, 110–114.

Eele, A. and Richards, A. 2009. Path planning with avoidance using nonlinear branch-and-bound optimization. *Journal of Guidance, Control, and Dynamics*, **32**(2), 384–394.

Eun, Y. and Bang, H. 2006. Cooperative control of multiple unmanned aerial vehicles using the potential field theory. *Journal of Aircraft*, **43**, 1805–1814.

Fong, T., Thorpe, C. and Baur, C. 2003. Multi-robot remote driving with collaborative control. *IEEE Transactions on Industrial Electronics*, **50**(4), 699–704.

Fowler, J. M. and D'Andrea, R. A. 2003. Formation flight experiment. *IEEE Control Systems Magazine*, 35–43.

Gennery, D. B. 2004. Traversability analysis and path planning for a planetary rover. *Autonomous Robots*, 131–146.

Hebert, J., Jacques, D., Novy, M. and Pachter, M. 2001. Cooperative control of UAVs. *AIAA Guidance, Navigation, and Control Conf. and Exhibit*, Montreal. AIAA-2001-4240.

Howard, A., Parker, L. and Sukatme, G. 2006. Experiments with a large heterogeneous mobile robot team: exploration, mapping, deployment and detection. *International Journal of Robotics Research*, **25**(5–6), 431–447.

Iyengar, S., Jorgensen, C., Rao, N. and Weisbin, R. 1986. Robot navigation algorithm using learned spatial graphs. *Robotica*, **4**, 93–100.

Judd, K. and McLain, T. W. 2001. Spline based path planning for unmanned air vehicles. *AIAA Guidance, Navigation, and Control Conf. and Exhibit*, Montreal. AIAA-2001-4238.

Jun, M. and D'Andrea, R. 2003. Path planning for unmanned aerial vehicles in uncertain and adversarial environments. In *Cooperative Control: Models, Applications and Algorithms* (eds Butenko, R., Murphey, R. and Pardalos, P.), pp. 95–110. Kluwer Academic.

Jung, D. and Gupta, K. K. 1996. Octree-based hierarchical distance maps for collision detection. *Proc. IEEE Int. Conf. on Robotics and Automation*, vol. 1, pp. 454–459.

Kambhampari, L. S. and Davis, L. S. 1986. Multi-resolution path planning for mobile robots. *IEEE Journal of Robotics and Automation*, **2**(3), 135–145.

Kanayama, Y. and Hartman, B. I. 1989. Smooth local path planning for autonomous vehicles. *Proc. IEEE Int. Conf. on Robotics and Automation*, Scottsdale, AZ, pp. 1265–1270.

Kavraki, L. E., Svestka, P., Latombe, J.-C. and Overmars, M. 1996. Probabilistic roadmaps for path planning in high-dimensional configuration spaces. *IEEE Transactions on Robotics and Automation*, **12**, 566–580.

Khatib, O. 1985. Real time obstacle avoidance for manipulators and mobile robots. *Proc. IEEE Int. Conf. on Robotics and Automation*, pp. 500–505.

Komoriya, K. and Tanie, K. 1989. Trajectory design and control of a wheel-type mobile robot using B-spline curve. *IEEE–RSJ Int. Conf. on Intelligent Robots and Systems*, Tsukuba, pp. 398–405.

Koren, Y. and Borenstein, J. 1991. Potential field methods and their inherent limitations for mobile robot navigation. *Proc. IEEE Int. Conf. on Robotics and Automation*, pp. 1398–1404.

Latombe, J. C. 1991. *Robot Motion Planning*. Kluwer Academic.

Laumond, J. P. 1986. Feasible trajectories for mobile robots with kinematic and environment constraints. *Proc. Int. Conf. on Intelligent Autonomous Systems*, Amsterdam, pp. 346–354.

LaValle, S. 1998. Rapidly-exploring random trees: a new tool for path planning. Technical Report.

LaValle, S. M. 2006. *Planning Algorithm*. Cambridge University Press.

Lipschutz, M. 1969. *Schaum's Outline of Differential Geometry*. McGraw-Hill.

Liscano, R. and Green, D. 1989. Design and implementation of a trajectory generator for an indoor mobile robot. *IEEE–RSJ Int. Conf. on Intelligent Robots and Systems*, Tsukuba, pp. 380–385.

Massink, M. and Francesco, N. 2001. Modelling free flight with collision avoidance. *Proc. IEEE Int. Conf. on Engineering of Complex Computer Systems*, pp. 270–279.

McGee, T. G., Spry, S. and Hedrick, J. K. 2005. Optimal path planning in a constant wind with a bounded turning rate. *AIAA Guidance, Navigation, and Control Conf. and Exhibit*, San Francisco.

McLain, L. and Beard, R. W. 2000. Trajectory planning for coordinated rendezvous of unmanned air vehicles. *AIAA Guidance, Navigation, and Control Conf.*, Denver. AIAA-2000-4369.

Mitcheel, I. M. and Sastry, S. 2003. Continuous path planning with multiple constraints. *IEEE Conf. on Decision and Control*, pp. 5502–5507.

Mittal, S. and Deb, K. 2007. Three dimensional offline path planning for UAVs using multiobjective evolutionary algorithms. *Proc. IEEE Congress on Evolutionary Computation, CEC 2007*, Singapore, 25–28 September.

Murray, R. M. 2007. Recent research in cooperative control of multi-vehicle systems. *Journal of Dynamic Systems, Measurement and Control*, **129**(5), 571–583.

Nelson, W. L. 1989. Continuous curvature paths for autonomous vehicles. *IEEE Int. Conf. on Robotics and Automation*, Scottsdale, AZ, vol. 3, pp. 1260–1264.

Nikolos, I., Valavanis, K., Tsourveloudis, N. and Kostaras, A. 2003. Evolutionary algorithm based offline/online path planner for uav navigation. *IEEE Transactions on Systems, Man, and Cybernetics – Part B*, **33**, 898–912.

NMAB and ASEB (National Materials Advisory Board, and Aeronautics and Space Engineering Board) 2000. *Uninhabited Air Vehicles: Enabling Science for Military Systems*. National Academy Press.

Oliveira, P., Pascoal, A., Silva, V. and Silvestre, C. 1998. Mission control of the MARIUS AUV: system design, implementation, and sea trials. *International Journal of Systems Science*, **29**, 1065–1080.

OSD (Office of the Secretary of Defense) 2005. *Unmanned Aircraft Systems Roadmap 2005–2030*. http://ftp.fas.org/irp/program/collect/uav_roadmap2005.pdf

Patcher, M. and Chandler, P. R. 1998. Challenges of autonomous control. *IEEE Control Systems Magazine*, pp. 93–97.

Pettersson, P. O. and Doherty, P. 2004. Probabilistic roadmap based path planning for an autonomous unmanned aerial vehicle. In *Workshop on Connecting Planning and Theory with Practice* at *14th Int. Conf. on Automated Planning and Scheduling, ICAPS*.

Piazzi, A. and Bianco, C. L. 2000. Quintic G^2 splines for trajectory planning of autonomous vehicles. *IEEE Intelligent Vehicles Symp.*, Dearborn, MI, pp. 193–200.

Polymenakos, L. C., Bertsekas, D. P. and Tsitsiklis, J. N. 1998. Implementation of efficient algorithms for globally optimal trajectories. *IEEE Transactions on Automatic Control*, **43**, 278–283.

Qu, Z. 2008. *Cooperative Control of Dynamical Systems*. Springer.

Rabbath, C., Gagnon, E. and Lauzon, M. 2004. On the cooperative control of multiple unmanned aerial vehicles. *IEEE Canadian Review*, 8–15.

Reeds, J. A. and Shepp, R. A. 1990. Optimal paths for a car that goes both forward and backward. *Pacific Journal of Mathematics*, **145**(2), 367–393.

Ren, W., Beard, R. W. and McLain, T. W. 2004. Coordination variables and consensus building in multiple vehicle systems. In *Cooperative Control* (eds Kumar, V., Leonard, N. and Morse, A. S.), pp. 171–188. Springer.

Richards, A. and How, J. 2002. Aircraft trajectory planning with collision avoidance using mixed integer linear programming. *American Control Conf.*, pp. 1936–1941.

Richards, A., How, J., Schouwenaars, T. and Feron, E. 2001. Plume avoidance maneuver planning using mixed integer programming. *AIAA Guidance, Navigation, and Control Conf.*. AIAA-2001-4091.

Robb, M., White, B. A., Tsourdos, A. and Rulloda, D. 2005. Reachability guidance: a novel concept to improve mid-course guidance. *American Control Conf.*, pp. 339–345.

Roberts, G. C., Ramana, M. V., Corrigan, C., Kim, D. and Ramanathan, V. 2008. Simultaneous observations of aerosol–cloud–albedo interactions with three stacked unmanned aerial vehicles. *Proceedings of the National Academy of Sciences of the USA*, **105**(21), 7370–7375.

Sasiadek, J. and Duleba, I. 2000. 3D local trajectory planner for UAV. *Journal of Intelligent and Robotic Systems: Theory and Applications*, **29**, 191–210.

Scheuer, A. and Fraichard, T. 1997. Collision-free continuous curvature path planning for car-like robots. *IEEE Int. Conf. on Robotics and Automation*, Albuquerque, NM, pp. 867–873.

Schouwenaars, T., Moor, B., Feron, E. and How, J. 2001. Mixed integer programming for multi-vehicle path planning. *European Control Conf.*, Porto, Portugal.

Schouwenaars, T., Feron, E. and How, J. 2006. Multi-vehicle path planning for non-line of sight communication. *American Control Conf.*, Minnesota, 14–16 June.

Segovia, S., Rombaut, M., Preciado, A. and Meizel, D. 1991. Comparative study of the different methods of path generation for a mobile robot in a free environment. *Proc. Int. Conf. on Advanced Robotics*, pp. 1667–1670.

Shamma, J. S. (ed.) 2007. *Cooperative Control of Distributed Multi-Agent Systems*. John Wiley & Sons, Ltd, Chichester.

Shanmugavel, M., Tsourdos, A., Żbikowski, R. and White, B. A. 2005. Path planning of multiple UAVs using Dubins sets. *AIAA Guidance, Navigation, and Control Conf. and Exhibit*, San Francisco. AIAA-2005-5827.

Shanmugavel, M., Tsourdos, A., Żbikowski, R., White, B. A., Rabbath, C. A. and Lechevin, N. 2006a. A solution to simultaneous arrival of multiple UAVs using Pythagorean hodograph curves. *Proc. American Control Conf.*, Minneapolis, 14–16 June.

Shanmugavel, M., Tsourdos, A., Żbikowski, R. and White, B. A. 2006b. 3D Dubins sets based coordinated path planning for swarm of UAVs. *AIAA Guidance, Navigation, and Control Conf. and Exhibit*, Keystone, CO. AIAA-2006-6211.

Shanmugavel, M., Tsourdos, A., Żbikowski, R. and White, B. A. 2007a. Path planning of multiple UAVs with clothoid curves. *17th IFAC Symp. on Automatic Control in Aerospace*, Toulouse, France.

Shanmugavel, M., Tsourdos, A., Żbikowski, R. and White, B. A. 2007b. 3D path planning for multiple UAVs using Pythagorean hodograph curves. *AIAA Guidance, Navigation, and Control Conf. and Exhibit*, Hilton Head, SC.

Shanmugavel, M., Tsourdos, A., White, B. A. and Żbikowski, R. 2007c. Differential geometric path planning of multiple UAVs. *AMSE Journal of Dynamic Systems, Measurement and Control*, **129**(5), 620–632.

Shanmugavel, M., Tsourdos, A., White, B. A. and Żbikowski, R. 2009. Co-operative path planning of multiple UAVs using Dubins paths with clothoid arcs. *Control Engineering Practice*, **17**, in press.

Shima, T., Rasmussen, S. and Sparks, A. 2005. UAV cooperative multiple task assignments using genetic algorithms. *Proc. American Control Conf.*, Oregon, vol. 5, pp. 2989–2994.

Shkel, A. and Lumelsky, V. 1996. On calculation of optimal paths with constrained curvature: the case of long paths. *IEEE Int. Conf. on Robotics and Automation*, pp. 3578–3583.

Shladover, S. E. 1991. Automatic vehicle control developments in the PATH program. *IEEE Transactions on Vehicular Technology*, **40**, 114–130.

Smith, T. R., Hansmann, H. and Leonard, N. E. 2001. Orientation control of multiple underwater vehicles. *Proc. 40th IEEE Conf. on Decision and Control*, pp. 4598–4603.

Takahashi, A., Hongo, T. and Ninomiya, Y. 1989. Local path planning and control for AGV in positioning. *IEEE–RSJ Int. Conf. on Intelligent Robots and Systems*, Tsukuba, pp. 392–397.

Tompkins, P., Stentz, A. and Wettergreen, D. 2004. Global path planning for Mars rover exploration. *IEEE Aerospace Conf.*, vol. 2, pp. 801–815.

Tsitsiklis, J. N. 1995. Efficient algorithms for globally optimal trajectories. *IEEE Transactions on Automatic Control*, **40**, 1528–1538.

Uny Cao, Y., Fukunaga, A. and Kahng, A. 1997. Cooperative mobile robotics: antecedents and directions. *Autonomous Robots*, **4**, 7–27.

Vian, J. L. and More, J. R. 1989. Trajectory optimization with risk minimisation for military aircraft. *AIAA Journal of Guidance, Control and Dynamics*, **12**, 311–317.

Wilson, J. 2007. UAV worldwide roundup 2007. *Aerospace America*, pp. 30–37.

Wu, X., Feng, Z., Zhu, J. and Allen, R. 2007. GA-based path planning for multiple AUVs. *International Journal of Control*, **80**(7), 1180–1185.

Yang, G. and Kapila, V. 2002. Optimal path planning for unmanned air vehicles with kinematic and tactical constraints. *IEEE Conf. on Decision and Control*, pp. 1301–1306.

Yuh, J. 1995. Developments in underwater robotics. *Proc. IEEE Conf. on Robotics and Automation*, pp. 1862–1867.

Yuh, J. 2000. Underwater robotics. *Proc. IEEE Conf. on Robotics and Automation*, pp. 932–937.

Zabarankin, M., Uryasev, S. and Pardalos, P. 2002. Optimal risk path algorithms. In *Cooperative Control and Optimization* (eds Murphey, R. and Pardalos, P.), pp. 271–303. Kluwer Academic.

Zhang, D., Wang, L. and Yu, J. 2008. Geometric topology based cooperation for multiple robots in adversarial environments. *Control Engineering Practice*, **16**, 1092–1100.

Zheng, C., Li, L., Xu, F., Sun, F. and Ding, M. 2005. Evolutionary route planner for unmanned air vehicles. *IEEE Transactions on Robotics*, **21**, 609–620.

2

Path Planning in Two Dimensions

In order to produce flyable paths that connect selected points, the constraints of the UAV flight must be taken into account. The main constraint is that of limited curvature or turn radius. This, together with producing paths that are as short as possible, requires the development of approaches that will take these constraints into account. Quite often, it is advantageous to have flyable paths that are the shortest in order to minimise flight time as well as reducing fuel consumption, and the related advantage of minimum energy and power. The techniques can then be extended to produce paths with longer lengths for the purpose of obstacle avoidance or to synchronise the arrival times for autonomous vehicles. Algorithms for the design of shortest routes are treated in various fields such as computational geometry, operations research and logistics. One well-researched problem in computational geometry is that of finding the shortest route in a given graph (Fredman and Tarjan 1987). The travelling salesman problem (TSP) and the Chinese postman problem (CPP) are widely studied in the area of operations research, but these look at shortest paths in an existing graph and the solutions do not have vehicle constraints built into their solution.

This chapter discusses the design of flyable paths, which are either polynomial curves, or entails the inclusion of lines and arcs to compute paths between the initial and final poses. There are a wide variety of such paths available in the literature. Three types of paths are studied here: (i) Dubins paths with constant-curvature arcs; (ii) clothoid paths, which are

Cooperative Path Planning of Unmanned Aerial Vehicles
Antonios Tsourdos, Brian White and Madhavan Shanmugavel
© 2011 John Wiley & Sons, Ltd

Dubins paths with variable-curvature arcs; and (iii) Pythagorean hodograph (PH) paths. The Dubins path is the shortest path, but has discontinuous curvature. The remaining two path types are continuous in curvature but are more complex to compute. In the first case, the constant-curvature arcs are replaced with clothoid arcs, which provide a smooth curvature transition between linear path segments. Finally, the PH curve is adopted from the literature for its rational properties in that the curve is parametrized by polynomials as a function of the path length and is designed as a single curve rather than as a set of joined segments.

2.1 Dubins Paths

In its simplest form, a path will consist of a combination of linear segments and constant-curvature segments. Combining both of these rectilinear and circular arcs will produce the shortest-length manoeuvre for a vehicle between two poses in space. In practice, this behaviour can be defined between two or more points in space in order to define the flight path for a UAV. In the case of an autonomous vehicle, in addition to the points, the orientation of the vehicle is also important. For example, it might be necessary to specify the angle of approach to an object for a better view. Therefore, it is important to find the shortest path between a starting position with orientation (starting pose) and a finishing position with orientation (finishing pose).

The shortest path connecting two poses is the Dubins path. A mathematical proof is provided by Dubins (1957). The simplest definition of a Dubins path is as follows: the shortest possible path that meets the maximum-curvature bound between two points with specific orientations in a plane is either a CLC or a CCC path, or a subset of them, where C represents circular arc and L represents straight-line tangent to C. This is further proved by other methods. Optimal control is used in Boissonnat *et al.* (1994) and Sussmann and Tang (1991). A variation of a Dubins path that allows backward motion is studied in Reeds and Shepp (1990). In Shkel and Lumelsky (2001), the lengths of the CLC paths are calculated for a given set of poses. The CLC path is formed by connecting two circular arcs by a line that is tangential to them, and the CCC path is formed by three consecutive circular arcs that are tangential to each other. The subsets of these paths are CL, LC and CC. Figure 2.1 shows the CLC and CCC paths. An application that involves tracking a target closely is studied using the CCC path in Wong *et al.* (2004).

We will focus on the CLC and CC paths, and, although there are a variety of methods to prove that the Dubins path is the shortest, the literature lacks

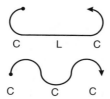

Figure 2.1   CLC and CCC types of Dubins path

simple methods of producing the Dubins path. Considering this, we focus on producing a Dubins path using the principles of both analytical and differential geometries.

2.2 Designing Dubins Paths using Analytical Geometry

Analytical geometry is chosen first as it is a familiar approach that will give some insight into the practical production of Dubins paths. From the previous section, the Dubins path is formed by a choice of two tangents connecting the two circular arcs. The initial and final positions lie on the arcs, the radii of the arcs are defined by the radii of curvature obtained from the turning radii of the vehicles, and the centres of the arcs are defined by the centres of curvature. Hence, the problem is reduced to finding the common tangents between two circular arcs. As stated previously, there are two common tangents between two circles: (i) internal/inner tangent, and (ii) external/outer tangent. The first is parallel to the line connecting the centres of the arcs for the same radii and the second one crosses the centreline. If the initial and final arc circles intersect, it is possible only to have the external tangent solution, as the internal tangent does not exist. This will be explained later in the chapter.

As we are interested in the poses rather than waypoints, there are four possible solutions. Figure 2.2 shows that, for a given pose, the UAV can either turn right or turn left. Hence, the path can start and finish with either clockwise (right) or anticlockwise (left) turns. The paths with external tangents follow a right-to-right turn or a left-to-left turn, while the paths with internal tangents follows either a right-then-left or a left-then-right turn. The figure shows that, as there are two turns for a given pose, there are two tangent circles, one with right turn designated R on the arc C_1 and the other with left turn designated L on the arc C_2. Hence, a pair of poses produces four Dubins paths.

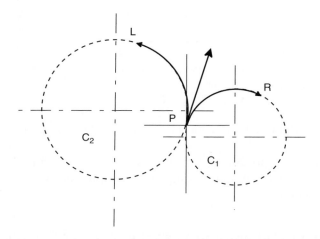

Figure 2.2   Tangent circles

Figures 2.3 and 2.4 respectively show the four Dubins paths with both external and internal tangents. Designing a Dubins path requires the following input parameters.

- (i) Start pose: $P_s(x_s, y_s, \phi_s)$.
- (ii) Finish pose: $P_f(x_f, y_f, \phi_f)$.
- (iii) Start curvature: $\kappa_s \leq \kappa_{max}$, where $\kappa_s = 1, \rho_s$ and ρ_s is the radius of curvature.
- (iv) Finish curvature: $\kappa_f \leq \kappa_{max}$, where $\kappa_f = 1/\rho_f$ and ρ_f is the radius of curvature,

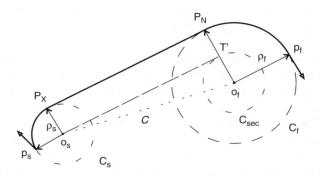

Figure 2.3   Dubins path with external tangent. Reprinted with permission of ASME

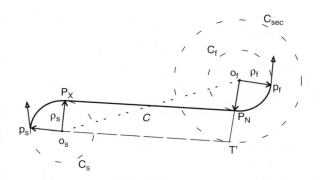

Figure 2.4 Dubins path with internal tangent

Here the suffix 's' represents the start parameters and the suffix 'f' represents the final parameters. Using the definition of path planning in equation (1.7) in Chapter 1, the design of the Dubins path can mathematically be represented by

$$P_s(x_s, y_s, \phi_s) \xrightarrow{r(q)} P_f(x_f, y_f, \phi_f), \qquad |\kappa(t)| \leq \kappa_{max}. \tag{2.1}$$

The solution to equation (2.1) produces a set of Dubins paths. The solution involves the steps in the following subsections.

2.2.1 Dubins Path: External Tangent Solution

The solution using the external tangent is obtained as follows.

1. Find the centre $o_s(x_{cs}, y_{cs})$ of the start circle C_s and the centre $o_s(x_{cf}, y_{cf})$ of the finish circle C_f using

$$x_{cs} = x_s - \frac{1}{\kappa_s} \cos(\phi_s \pm \pi/2), \tag{2.2a}$$

$$y_{cs} = y_s - \frac{1}{\kappa_s} \sin(\phi_s \pm \pi/2), \tag{2.2b}$$

$$x_{cf} = x_f - \frac{1}{\kappa_f} \cos(\phi_f \pm \pi/2), \tag{2.2c}$$

$$y_{cf} = y_f - \frac{1}{\kappa_f} \sin(\phi_f \pm \pi/2), \tag{2.2d}$$

where C_s and C_f are also called primary circles.

2. Draw a secondary circle C_{sec} of radius $|1/\kappa_f - 1/\kappa_s|$ at O_f for $1/\kappa_f \leq 1/\kappa_s$.
3. Connect the centres O_s and O_f by a line c, called the centreline. The length of the centreline is $|c| = \sqrt{(x_{cs} - x_{cf})^2 + (y_{cs} - y_{cf})^2}$.
4. Draw a perpendicular to c at O_f. The perpendicular intersects C_{sec} at T' and C_f at P_N. The point P_N is called the tangent entry point.
5. Connect O_s with T'.
6. Draw a line from O_s parallel to $O_f P_N$ which meets C_s at P_X, called the tangent exit point.
7. Connect the points P_X and P_N which is parallel to the line $O_s T'$. The line $P_X P_N$ is the external tangent.
8. Draw an arc of radius ρ_s from P_s to P_X, a line from P_X to P_N, and another arc of radius ρ_f from P_N to P_f. The connected arcs and the line form the Dubins path with external tangent.

From the resulting diagram, the Dubins solution can be obtained. In Figure 2.3 the triangle $\triangle O_s O_f T'$ is a right-angled triangle with sides $O_f T'$ and $O_s T'$, and hypotenuse $O_s O_f$, where $\|O_s T'\| = |\rho_f - \rho_s|$. Next calculate the angle between $O_s O_f$ and $O_s T'$ as

$$\alpha = \arcsin\left(\frac{\rho_f - \rho_s}{|c|}\right). \tag{2.3}$$

The slope of the centreline is

$$\beta = \arctan\left(\frac{y_{cf} - y_{cs}}{x_{cf} - x_{cs}}\right), \tag{2.4}$$

so the tangent exit point $P_X(x_{P_X}, y_{P_X})$ on C_s and entry point $P_N(x_{P_N}, y_{P_N})$ on C_f can be calculated as

$$x_{P_X} = x_{cs} + \frac{1}{\kappa_s}\cos(\phi), \tag{2.5a}$$

$$y_{P_X} = y_{cs} + \frac{1}{\kappa_s}\sin(\phi), \tag{2.5b}$$

$$x_{P_N} = x_{cf} + \frac{1}{\kappa_f}\cos(\phi), \tag{2.5c}$$

$$y_{P_N} = y_{cf} + \frac{1}{\kappa_f}\sin(\phi), \tag{2.5d}$$

Table 2.1 Calculation of tangent exit and entry points for external tangent

	Start turn	Finish turn
ϕ_{right}	$\alpha + \beta + \dfrac{\pi}{2}$	$\alpha + \beta + \dfrac{\pi}{2}$
ϕ_{left}	$\beta - \alpha + \dfrac{3\pi}{2}$	$\beta - \alpha + \dfrac{3\pi}{2}$

For $\kappa_s = \kappa_f$, α will vanish, which gives $O_s O_f$ parallel to $P_X P_N$.

where ϕ is obtained from Table 2.1. The arc angles which rotate from the initial pose to the tangent pose and from the tangent pose to the finish pose are then given by the following:

1. rotation from the start point to the tangent exit point P_X by ($\psi_s = \phi_s \pm \frac{1}{2}\pi + \phi$);
2. zero rotation from P_X to P_N; and
3. rotation from the tangent entry point to P_N by ($\psi_f = \phi_f \pm \frac{1}{2}\pi + \phi$).

The straight-line segment is finally calculated by using Pythagoras's theorem on the tangent entry and exit points:

$$l = \sqrt{(x_{P_X} - x_{P_N})^2 + (y_{P_X} - y_{P_N})^2}. \tag{2.6}$$

2.2.2 Dubins Path: Internal Tangent Solution

The design of the Dubins path using the internal tangent is similar to that using the external tangent, except that it differs only in the angle of rotation. The steps are as follows.

1. Find the centre $o_s(x_{cs}, y_{cs})$ of the start circle C_s and the centre $o_s(x_{cf}, y_{cf})$ of the finish circle C_f using

$$x_{cs} = x_s - \frac{1}{\kappa_s} \cos(\phi_s \pm \pi/2), \tag{2.7a}$$

$$y_{cs} = y_s - \frac{1}{\kappa_s} \sin(\phi_s \pm \pi/2). \tag{2.7b}$$

$$x_{cf} = x_f - \frac{1}{\kappa_f}\cos(\phi_f \pm \pi/2), \qquad (2.7c)$$

$$y_{cf} = y_f - \frac{1}{\kappa_f}\sin(\phi_f \pm \pi/2), \qquad (2.7d)$$

where C_s and C_f are also called primary circles.

2. Draw a secondary circle C_{sec} of radius $|1/\kappa_f + 1/\kappa_s|$ at o_f for $1/\kappa_f \le 1/\kappa_s$.
3. Connect the centres o_s and o_f by a line c, called the centreline. The length of the centreline is $|c| = \sqrt{(x_{cs} - x_{cf})^2 + (y_{cs} - y_{cf})^2}$.
4. Draw a perpendicular to c at o_f. The perpendicular intersects C_{sec} at T' and C_f at P_N. The point P_N is called the tangent entry point.
5. Connect o_s with T'.
6. Draw a line from o_s parallel to $o_f P_N$ which meets C_s at P_X, called the tangent exit point.
7. Connect the points P_X and P_N which is parallel to the line $o_s T'$. The line $P_X P_N$ is the internal tangent.
8. Draw an arc of radius ρ_s from P_s to P_X, a line from P_X to P_N, and another arc of radius ρ_f from P_N to P_f. The connected arcs and the line form the Dubins path with internal tangent.

The triangle $\triangle o_s o_f T'$ in Figure 2.3 is a right-angled triangle with sides $o_f T'$ and $o_s T'$, and hypotenuse $o_s o_f$, where $\|o_s T'\| = |\rho_f + \rho_s|$. Next calculate the angle between $O_s O_f$ and $O_s T'$ as

$$\alpha = \arcsin\left(\frac{\rho_f + \rho_s}{|c|}\right). \qquad (2.8)$$

The slope of the centreline is

$$\beta = \arctan\left(\frac{y_{cf} - y_{cs}}{x_{cf} - x_{cs}}\right), \qquad (2.9)$$

so the tangent exit point $P_X(x_{P_X}, y_{P_X})$ on C_s and entry point $P_N(x_{P_N}, y_{P_N})$ on C_f can be calculated as

$$x_{P_X} = x_{cs} + \frac{1}{\kappa_s}\cos(\phi), \qquad (2.10a)$$

$$y_{P_X} = y_{cs} + \frac{1}{\kappa_s}\sin(\phi), \qquad (2.10b)$$

$$x_{P_N} = x_{cf} + \frac{1}{\kappa_f}\cos(\phi), \qquad (2.10c)$$

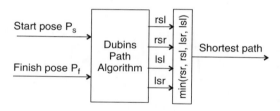

Figure 2.5 Block diagram of path planner to generate the shortest flyable paths

$$y_{PN} = y_{cf} + \frac{1}{\kappa_f} \sin(\phi), \qquad (2.10d)$$

where ϕ is obtained from Table 2.1. The angle of rotations required to design the Dubins path are then as follows:

1. rotation from the start point to P_X by ($\psi_s = \phi_s \pm \frac{1}{2}\pi + \phi$);
2. zero rotation from P_X to P_N; and
3. rotation from the tangent exit point to P_N by ($\psi_f = \phi_f \pm \frac{1}{2}\pi + \phi$).

Figure 2.5 shows the mechanism for producing the Dubins paths. By inputting the start and finish poses, P_s and P_f, and the start and finish curvatures, κ_s and κ_f, the algorithm produces the four possible paths. The shortest path can then be chosen from the set of paths. Sometimes, applications may need a UAV to reach a point irrespective of the approach direction. In such cases, the final arc can be deleted and the straight-line segment terminates on the final position. For the other case, the final orientation is set, but the approach direction is taken as a free variable. For this case, there are eight Dubins paths possible, as shown in Figure 2.6. If both the start and finish orientations are free, the number of paths will increase to 16.

2.3 Existence of Dubins Paths

The design of Dubins paths involves finding the tangent $P_X P_N$ points between the two circular arcs. If it is not possible to construct the tangent points, then the Dubins path does not exist. However, if these two tangent points coincide, the straight-line segment has zero length, which results in a CC path, that is, two circular arcs connected tangentially. The external tangent vanishes when one turning circle lies entirely within the other, while the

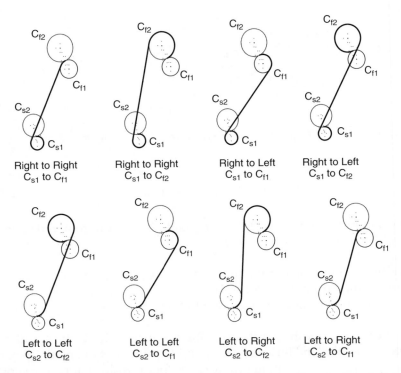

Figure 2.6 Dubins paths with ϕ_f as a free variable. The start turn is either clockwise or anticlockwise. Four possible turns on each tangent circle produce eight paths. Reprinted with permission of the American Institute of Aeronautics and Astronautics

internal tangent vanishes when the tangent circles intersect. In terms of the turning radii ρ_s and ρ_f and the central distance $|c|$ in equation (2.11), the existence of the Dubins path is given by the conditions

$$\text{external tangent,} \qquad |\rho_f - \rho_s| < |c|, \qquad (2.11a)$$

$$\text{internal tangent,} \qquad |\rho_s + \rho_f| < |c|. \qquad (2.11b)$$

However, if a solution is not possible for a specific start and finish arc curvature, a scaled version of the path can be produced by increasing the curvatures of either the start, finish or both arcs. From equation (2.11), it can be said that the existence of the Dubins path is roughly a function of its turning radii or curvatures.

Table 2.2 Calculation of tangent exit and entry points for internal tangent

	Start turn	Finish turn
ϕ_{right}	$\alpha + \beta$	$\beta - \alpha + \pi$
ϕ_{left}	$\beta - \alpha + 2\pi$	$\beta + \alpha + \pi$

For $\kappa_s = \kappa_f$, α will vanish, which gives $O_s O_f$ parallel to $P_X P_N$.

2.4 Length of Dubins Path

The Dubins path is a composite path made of two circular arcs and a straight line, and the length of the Dubins path is the sum of their individual lengths, that is

$$s_{Dubins} = s_{arc,\,start} + s_{tangent} + s_{arc,\,finish}. \tag{2.12}$$

For the case of design using the analytic geometry, we have

$$s_{Dubins} = \frac{1}{\kappa_s} \phi_s + s_t + \frac{1}{\kappa_f} \phi_f \tag{2.13a}$$

$$= f(\kappa_s, \kappa_f), \tag{2.13b}$$

where s_{Dubins} is the length of the Dubins path, ϕ_s and ϕ_f respectively are the start and finish arc angles obtained from Tables 2.1 and 2.2, and $s_t = \|P_X P_N\|$. From equation (2.13b), the path length is simply a function of the start and finish radii. However, this function is not a simple equation and, by varying the values of the curvatures, the length of the tangent can be varied.

2.5 Design of Dubins Paths using Principles of Differential Geometry

In this approach, the pose is represented by vectors of unit length centred at the start and finish positions. The pose $P(x, y, \theta)$ is considered as a vector with its origin at position (x, y) with orientation θ. This vector is tangential to the path, and represents the velocity vector of the UAV. The problem is simplified to finding one or more flyable paths that connect the two start and finish velocity vectors. For path planning in 2D, the initial and final tangent vectors are coplanar, hence the initial and final turning circles

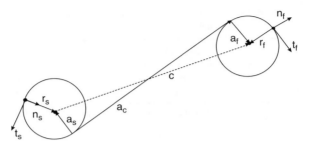

Figure 2.7 Dubins arc geometry. Reprinted with permission of the American Institute of Aeronautics and Astronautics

and the connecting tangent lie in the same plane. The derivation starts by assigning a basis frame for each path segment of the Dubins path. More details defining differential geometry are given in Appendix A.

A 2D Dubins path is shown in Figure 2.7. The Dubins path is assigned with three frames for each manoeuvre: two turning manoeuvres, and a straight-line manoeuvre. Hence the start frame e_s, the finish frame e_f and the connecting frame e_c are defined. Each basis frame is thus defined such that $e = [t \; n]$, where t is the unit tangent, n is the unit normal and the subscripts 's', 'f' and 'c' respectively are the start, finish and connecting frames. The components of the frame are the unit tangent vector, which is parallel to the velocity vector, and the unit normal vector, perpendicular to the tangent vector.

Figure 2.7 shows the start and finish basis vectors. Once the frames are established, the start and finish turn radius vectors r_s and r_f can be directly written in their respective frames e_s and e_f. The sign of the initial and final manoeuvre can be determined by designating either a left or right turn, which translates into a position or negative turn radius vector. Viewed from each position, a positive or negative rotation will also define the sign of the curvature for each manoeuvre. Hence each vector can be defined as follows:

$$r_s = e_s \begin{pmatrix} 0 \\ \pm 1/\kappa_s \end{pmatrix}, \qquad e_s = [t_s \; n_s], \tag{2.14}$$

$$r_f = e_f \begin{pmatrix} 0 \\ \pm 1/\kappa_f \end{pmatrix}, \qquad e_f = [t_f \; n_f], \tag{2.15}$$

$$a_c = e_c \begin{pmatrix} a \\ 0 \end{pmatrix}, \qquad e_c = [t_c \; n_c]. \tag{2.16}$$

Here κ_s is the curvature of the initial manoeuvre, κ_f is the curvature of the final manoeuvre, and the curvature of the connecting tangent vector is zero with length a.

Now, the problem of designing the Dubins path is finding the angle of rotation of the tangent vector t_s to the finish tangent vector t_f via the connecting vector a_c, so that the start frame coincides with the finish frame. These manoeuvre (and hence velocity) vectors t_s and t_f are related by

$$t_f = \mathbf{R}(\phi)t_s, \tag{2.17}$$

where $\mathbf{R}(\phi)$ is the rotation matrix required to change the axis set from initial to final axes. Hence, we have

$$\cos(\phi) = t_f' t_s, \tag{2.18}$$

from which the total rotation angle ϕ can be determined. The connecting vectors a_s, a_f and a_c form an orthogonal set of vectors. In order to determine the vectors, first define the connecting vector a_c as

$$t_c = \mathbf{R}(\phi_s)t_s, \tag{2.19}$$

where t_c is the basis vector defining the connecting vector and ϕ_s is the arc angle of the starting path segment. If the position of the final point p_f relative to the start position p_s is measured in start basis axes e_s, we have

$$p_f - p_s = e_s p,$$
$$p = \begin{pmatrix} p_t \\ p_n \end{pmatrix}. \tag{2.20}$$

Hence, the vector sum for the position vector in start axes is given by

$$p = r_s - a_s + a_c + a_f - r_f,$$
$$p - r_s + r_f = -a_s + a_c + a_f. \tag{2.21}$$

The left-hand side of this equation represents the vector connecting the centres of the turn circles. Hence

$$ct_c = -a_s + a_c + a_f, \tag{2.22}$$

where c is the length of the centre vector.

The remaining connecting vectors a_s, a_f and a_c can be written in terms of the start basis vectors as

$$a_s = \mathbf{R}(\phi_s)' \begin{pmatrix} 0 \\ \pm 1/\kappa_s \end{pmatrix},$$

$$a_f = \mathbf{R}(\phi_s)' \begin{pmatrix} 0 \\ \pm 1/\kappa_f \end{pmatrix}, \tag{2.23}$$

$$a_c = \mathbf{R}(\phi_s)' \begin{pmatrix} a \\ 0 \end{pmatrix}.$$

The centre vector in equation (2.22) now becomes

$$ct_c = -\mathbf{R}(\phi_s)' \begin{pmatrix} 0 \\ \pm 1/\kappa_s \end{pmatrix} + \mathbf{R}(\phi_s)' \begin{pmatrix} a \\ 0 \end{pmatrix} + \mathbf{R}(\phi_s)' \begin{pmatrix} 0 \\ \pm 1/\kappa_f \end{pmatrix}$$

$$= \mathbf{R}(\phi_s)' \begin{pmatrix} a \\ \pm 1/\kappa_f - \pm 1/\kappa_s \end{pmatrix}. \tag{2.24}$$

This is a rotation equation, hence the right-hand vector must have the same magnitude as the left, to give

$$\left| \frac{1}{c} \begin{pmatrix} a \\ \pm 1/\kappa_f - \pm 1/\kappa_s \end{pmatrix} \right| = 1 \tag{2.25}$$

or

$$\left(\frac{a}{c} \right)^2 + \frac{1}{c^2} \left(\frac{\pm 1}{\kappa_f} - \frac{\pm 1}{\kappa_s} \right)^2 = 1$$

$$\left(\frac{a}{c} \right)^2 = 1 - \frac{1}{c^2} \left(\frac{\pm 1}{\kappa_f} - \frac{\pm 1}{\kappa_s} \right)^2. \tag{2.26}$$

This can be used to test for a feasible solution, by

$$1 - \frac{1}{c^2} \left(\frac{\pm 1}{\kappa_f} - \frac{\pm 1}{\kappa_s} \right)^2 > 0. \tag{2.27}$$

This condition is used to test for the existence of a feasible solution. Also, this condition is equivalent to the conditions in section 2.3.

In order to compute the rotation angle ϕ_s, the equation can be written in the form

$$t_c = \mathbf{R}(\phi_s)' \begin{pmatrix} (1/c)\sqrt{c^2 - (\pm 1/\kappa_f - \pm 1/\kappa_s)^2} \\ (1/c)(\pm 1/\kappa_f - \pm 1/\kappa_s) \end{pmatrix},$$

$$\mathbf{R}(\phi_s) = \begin{pmatrix} \cos(\phi_s) & -\sin(\phi_s) \\ \sin(\phi_s) & \cos(\phi_s) \end{pmatrix}. \tag{2.28}$$

Solving for ϕ_s gives

$$\begin{pmatrix} \cos(\phi_s) \\ \sin(\phi_s) \end{pmatrix} = \mathbf{R}(c, \kappa_s, \kappa_f)t_c, \tag{2.29}$$

where

$$\mathbf{R}(c, \kappa_s, \kappa_f) = \frac{1}{c} \begin{pmatrix} \sqrt{c^2 - (\pm 1/\kappa_f - \pm 1/\kappa_s)^2} & -(\pm 1/\kappa_f - \pm 1/\kappa_s) \\ (\pm 1/\kappa_f - \pm 1/\kappa_s) & \sqrt{c^2 - (\pm 1/\kappa_f - \pm 1/\kappa_s)^2} \end{pmatrix}. \tag{2.30}$$

As the total rotation angle for the path has been determined from equation (2.18), the final arc angle ϕ_f can then be calculated using

$$\phi = \phi_s + \phi_f,$$

$$\phi_f = \phi - \phi_s. \tag{2.31}$$

2.5.1 Dubins Path Length

For the case of design using differential geometry, the corresponding equations (2.13) become

$$h_{\text{Dubins}} = \frac{\phi_s}{\kappa_s} + a + \frac{\phi_f}{\kappa_f} \tag{2.32}$$

$$= h(\kappa_s, \kappa_f). \tag{2.33}$$

Using the differential geometric solution for a range of curvatures, the resulting set of Dubins paths are shown in Figure 2.8. These show that the paths start and finish at the same poses but the paths have different lengths. The lengths of the solutions are shown in Figure 2.9. Figure 2.5 is also relevant

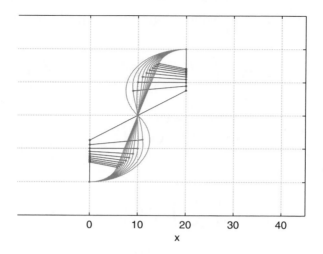

Figure 2.8   Set of Dubins paths over a range of κ

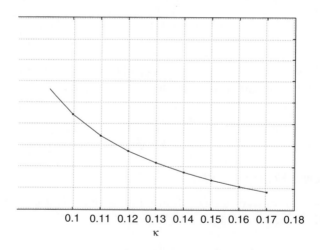

Figure 2.9   Set of Dubins path lengths over a range of κ

to this algorithm for the calculation of the shortest path. The algorithm block calculates the possible path for the input poses.

From the previous two sections, the important point to notice is that the following results are identical: (i) existence of path in equations (2.27) and (2.11), and (ii) path length in equations (2.33) and (2.13a).

2.6 Paths of Continuous Curvature

The Dubins path has a simple geometry and hence is easy to produce and implement. However, changes in path curvature are discontinuous when the path changes from an arc to a straight-line segment and vice versa. This discontinuity can prove difficult for practical implementation purposes, in that a UAV will have to follow such a path and it cannot instantaneously change its behaviour as it crosses the segment boundary. A change in curvature can be directly related to a change in lateral acceleration. In fact, for a UAV travelling at a constant speed v m s^{-1}, the curvature of the path that it is following and the acceleration normal to the path (the lateral acceleration) are given by

$$a = v^2 \kappa, \tag{2.34}$$

where a m s^{-2} is the lateral acceleration of the UAV and κ is the curvature of the path (see equation (1.6)). Hence a sudden change in acceleration demand will occur as the UAV crosses the path segment boundaries, which will produce errors in path following. Hence, it is important for the UAV to follow paths that are not discontinuous in curvature. From the principles of differential geometry, the second derivative of a curve with respect to a path parameter q represents acceleration. (Note that the path parameter can be any convenient parameter, such as the distance along the path or another convenient parameter such as a rotation angle for a circle – see Appendix A.) Hence, smooth motion requires at least non-vanishing first two derivatives.

Geometrically, a path is completely determined by its curvature, and curvature is proportional to lateral acceleration for a constant-speed UAV. The curvature $\kappa(q)$ of a curve $r(q) = (x(q), y(q))$, where q is defined as the curve parameter, is

$$\kappa(q) = \frac{\dot{r} \times \ddot{r}}{|\dot{r}|^3} \tag{2.35}$$

$$= \frac{\dot{x}(q)\ddot{y}(q) - \dot{y}(q)\ddot{x}(q)}{(\sqrt{\dot{x}^2 + \dot{y}^2})^3}, \tag{2.36}$$

where the derivatives $\dot{x} = dx/dq$, $\dot{y} = dy/dq$, $\ddot{x} = d^2x/dq^2$ and $\ddot{y} = d^2y/dq^2$ are defined relative to the path parameter.

From equation (2.36), the curvature is a function of the first two derivatives of a curve, so the path must be at least twice continuously differentiable; that is, it must have C^2 continuity. In the Dubins path, as shown in Figure 2.10,

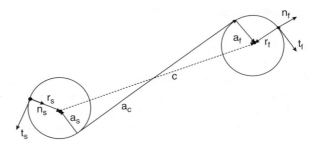

Figure 2.10   Dubins path. Reprinted with permission of the American Institute of Aeronautics and Astronautics, & ASME

the arcs have constant curvature and the line has zero curvature, but the segment boundary has a discontinuity in curvature. However, the curve does not have a sudden change in direction, which means that the path has a smooth first derivative but a discontinuous second derivative. Hence the Dubins path has C^1 continuity but not a C^2 continuity. (The notation C^2 represents continuity up to and including the second derivative, and C^1 represents continuity up to and including the first derivative.)

A path of continuous curvature can be produced either by a composite path formed by joining segments that have equal curvature at their boundaries, or by a producing a single curve with C^2 continuity. The first option can be met by using clothoid arcs, which provide the required curvature continuity across segment boundaries. The second option is met by using compact, well-defined paths using Pythagorean hodographs, which is a single path rather than a collection of path segments.

2.7 Producing Flyable Clothoid Paths

A clothoid path has a property that its curvature varies linearly over the path length. The curvature profile of such a path, compared to a Dubins path, is shown in Figure 2.11. The figure shows that the constant-curvature arc of the Dubins path is replaced by a triangular curvature profile, which varies linearly along the arc length, going from zero at the boundaries to a maximum at the arc mid-point. This linear variation of curvature with path length of the clothoid enables a smooth transition to and from the zero-curvature line segment.

The method of producing the composite clothoid path is similar to that of the Dubins path (as detailed in section 2.5). The only difference is that

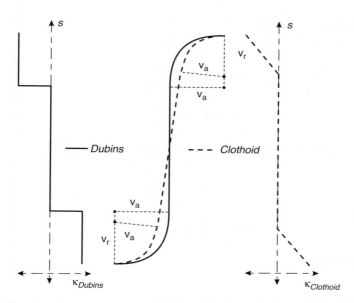

Figure 2.11 Curvature profiles of Dubins and clothoid paths. Reproduced by permission of Elsevier

the clothoid arc segment is no longer the arc of a circle and is now defined by calculating Fresnel integrals. For a clothoid arc, the path direction is no longer a simple function of the path variable q but must be computed by integrating the curvature along the path length to get

$$\phi(q) = \int_0^q \kappa \frac{k}{h} \, dk$$

$$= \frac{\kappa}{2h} q^2, \tag{2.37}$$

where q is the path distance variable along the arc, h is the total arc length and κ is the curvature at h. The position vector of the end point is given by the x and y positions. These are obtained by integration:

$$x(h) = \int_0^h \cos(\phi) \, dq,$$

$$y(h) = \int_0^h \sin(\phi) \, dq. \tag{2.38}$$

From equation (2.37), the angle $\phi(q)$ through which the trajectory moves over the total arc length h can be used to give

$$x(h) = \int_0^h \cos\left(\frac{\kappa}{2h}q^2\right) dq,$$

$$y(h) = \int_0^h \sin\left(\frac{\kappa}{2h}q^2\right) dq. \qquad (2.39)$$

These integrals are scaled Fresnel integrals and are given by

$$C(h) = \int_0^h \cos\left(\frac{\kappa}{2h}q^2\right) dq,$$

$$S(h) = \int_0^h \sin\left(\frac{\kappa}{2h}q^2\right) dq. \qquad (2.40)$$

Hence we have

$$x(h) = C(h),$$

$$y(h) = S(h). \qquad (2.41)$$

The integrals can be evaluated more easily by a change of variable given by

$$\bar{q} = \sqrt{\frac{\kappa}{2h}}q,$$

$$\bar{h} = \sqrt{\frac{\kappa}{2h}}h = \sqrt{\frac{\kappa h}{2}}.$$

Hence

$$dq = \sqrt{\frac{2h}{\kappa}}\, d\bar{q} \qquad (2.42)$$

and the integrals can be rewritten in the form

$$C(h) = \sqrt{\frac{2h}{\kappa}} \int_0^{\bar{h}} \cos((\bar{q})^2)\, d\bar{q}, \qquad (2.43a)$$

$$S(h) = \sqrt{\frac{2h}{\kappa}} \int_0^{\bar{h}} \sin((\bar{q})^2)\, d\bar{q}. \qquad (2.43b)$$

Figure 2.12 shows the clothoid trajectory, where the arcs are now not simple circles. The $x(h)$ and $y(h)$ positions can be constructed from the basis vectors

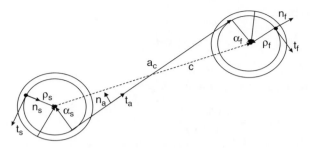

Figure 2.12 Path with clothoid arc geometry. Reproduced by permission of Elsevier

defined in the same manner as the Dubins path for the start pose (t_s, n_s), the straight-line segment (t_a, n_a) and the finish segment (t_f, n_f), to give

$$p = \rho n_s + \alpha n_a, \qquad (2.44)$$

where ρ and α are the lengths of the two vectors. These lengths can be calculated by noting that, as both n_s and n_a are basis vectors, they are of unit length. From Figure 2.12, we have

$$n_s = \begin{pmatrix} 0 \\ 1 \end{pmatrix},$$

$$n_a = \begin{pmatrix} \sin(\phi_s) \\ -\cos(\phi_s) \end{pmatrix}. \qquad (2.45)$$

Hence, we have

$$p = \begin{pmatrix} C(h) \\ S(h) \end{pmatrix}$$

$$= \rho \begin{pmatrix} 0 \\ 1 \end{pmatrix} + \alpha \begin{pmatrix} \sin(\phi_s) \\ -\cos(\phi_s) \end{pmatrix}. \qquad (2.46)$$

This gives

$$\alpha = \frac{C(h)}{\sin(\phi)},$$

$$\rho = \left(S(h) + \frac{1}{\tan(\phi)} C(h) \right). \qquad (2.47)$$

Now we also have

$$\phi_s = \phi(h) = \frac{\kappa}{2}h. \tag{2.48}$$

Converting to angles using $h = 2\phi_s/\kappa$ gives

$$\alpha = \frac{C(2\phi_s/\kappa)}{\sin(\phi_s)},$$

$$\rho = \left[S\left(\frac{2\phi_s}{\kappa}\right) + \frac{1}{\tan(\phi_s)}C\left(\frac{2\phi_s}{\kappa}\right) \right], \tag{2.49}$$

where

$$C\left(\frac{2\phi_s}{\kappa}\right) = \frac{2}{\kappa}\sqrt{\phi_s}\int_0^{\sqrt{\phi_s}} \cos((\bar{q})^2)\, d\bar{q}, \tag{2.50a}$$

$$S\left(\frac{2\phi_s}{\kappa}\right) = \frac{2}{\kappa}\sqrt{\phi_s}\int_0^{\sqrt{\phi_s}} \sin((\bar{q})^2)\, d\bar{q}. \tag{2.50b}$$

This implies that there is no closed-form solution to the clothoid trajectory, and iteration is required for a solution. For a two-dimensional manoeuvre, the initial and final tangent vectors are coplanar and the straight-line manoeuvre is now not uniquely defined for this case and so must be calculated. The 2D clothoid arc is shown in Figure 2.12.

The figure shows two circles of radius ρ and γ. Also, from the figure, the sign of the manoeuvre can be determined by considering the centreline between the two positions. Viewed from each position, a positive or negative rotation from the tangent vector to the centre vector will define the sign of the curvature for each manoeuvre. Also, from the figure, we have

$$r_s = e_s \begin{pmatrix} 0 \\ \pm\rho_s \end{pmatrix},$$

$$e_s = \begin{bmatrix} t_s & n_s \end{bmatrix}, \tag{2.51}$$

where α_s and ρ_s are the start and finish radii of the initial manoeuvre, given by

$$\alpha_s = \frac{C(2\phi_s/\kappa)}{\sin(\phi_s)},$$

$$\rho_s = \left[S\left(\frac{2\phi_s}{\kappa}\right) + \frac{1}{\tan(\phi_s)}C\left(\frac{2\phi_s}{\kappa}\right) \right]. \tag{2.52}$$

Similarly,

$$r_f = e_f \begin{pmatrix} 0 \\ \pm\rho_f \end{pmatrix},$$

$$e_f = \begin{bmatrix} t_f & n_f \end{bmatrix}, \tag{2.53}$$

where ρ_f and α_f are the start and finish radii of the final manoeuvre, given by

$$\alpha_f = \frac{C(2\phi_f/\kappa)}{\sin(\phi_f)},$$

$$\rho_f = \left[S\left(\frac{2\phi_f}{\kappa}\right) + \frac{1}{\tan(\phi_f)} C\left(\frac{2\phi_f}{\kappa}\right) \right]. \tag{2.54}$$

The Frenet basis vectors are related by

$$e_f = \mathbf{R}(\phi_t)e_s, \tag{2.55}$$

where $\mathbf{R}(\phi_t)$ is the rotation matrix required to change the axis set from start to finish axes. Hence, we have

$$\phi_t = \phi_s + \phi_f,$$

$$\mathbf{R}(\phi_t) = e_s e_f'. \tag{2.56}$$

In a similar manner to the Dubins solution, the connecting vectors α_s, a_f and α_c form an orthogonal set of vectors. In order to determine the vectors, first define the connecting vector a_c in both initial and final axes, as

$$e_c = \mathbf{R}(\phi_s)e_s,$$

$$e_f = \mathbf{R}(\phi_f)e_c, \tag{2.57}$$

where e_c is the basis set defining the connecting vector. Hence, the total rotation matrix $\mathbf{R}(\phi_t)$ is given by

$$\mathbf{R}(\phi_t) = \mathbf{R}(\phi_f)\mathbf{R}(\phi_s). \tag{2.58}$$

If the position of the final point p_f relative to the start position p_s is measured in start axes e_s, we have

$$p_f - p_s = e_s p$$

$$p = \begin{pmatrix} p_t \\ p_n \end{pmatrix}. \tag{2.59}$$

Hence, the vector sum for the position vector in start axes is given by

$$p = \rho_s - \alpha_s + a_c + \alpha_f - \rho_f$$

$$p - \rho_s + \rho_f = -\alpha_s + a_c + \alpha_f. \tag{2.60}$$

The left-hand side of this equation represents the vector connecting the centres of the turn circles:

$$c = p - \rho_s + \rho_f. \tag{2.61}$$

Hence we have

$$c = -\alpha_s + a_c + \alpha_f. \tag{2.62}$$

The centre vector c can be written in start axes, to give

$$c = c t_c = e_{ct} \begin{pmatrix} c \\ 0 \end{pmatrix},$$

$$e_{ct} = \begin{bmatrix} t_c & n_c \end{bmatrix}, \tag{2.63}$$

where e_{ct} is the basis vector set of the centre vector. The remaining connecting vectors α_s, α_f and a_c can be written in terms of the start basis vectors, as

$$\alpha_s = \mathbf{R}(\phi_s)' \begin{pmatrix} 0 \\ \pm\rho_s \end{pmatrix},$$

$$\alpha_f = \mathbf{R}(\phi_s)' \begin{pmatrix} 0 \\ \pm\rho_f \end{pmatrix}, \tag{2.64}$$

$$a_c = \mathbf{R}(\phi_s)' \begin{pmatrix} a \\ 0 \end{pmatrix}.$$

The centre vector (2.62) now becomes

$$c t_{ct} = -\mathbf{R}(\phi_s)' \begin{pmatrix} 0 \\ \pm\rho_s \end{pmatrix} + \mathbf{R}(\phi_s)' \begin{pmatrix} a \\ 0 \end{pmatrix} + \mathbf{R}(\phi_s)' \begin{pmatrix} 0 \\ \pm\rho_f \end{pmatrix}$$

$$= \mathbf{R}(\phi_s)' \begin{pmatrix} a \\ \pm\rho_f - \pm\rho_s \end{pmatrix}. \tag{2.65}$$

Normalizing the centre vector to unit magnitude gives

$$t_{ct} = \mathbf{R}(\phi_s)' \frac{1}{c} \begin{pmatrix} a \\ \pm \rho_f - \pm \rho_s \end{pmatrix}. \tag{2.66}$$

This is a rotation equation, which represents the rotation of a unit vector. Hence, the right-hand vector must have unit magnitude, to give

$$\left| \frac{1}{c} \begin{pmatrix} a \\ \pm \rho_f - \pm \rho_s \end{pmatrix} \right| = 1 \tag{2.67}$$

or

$$\left(\frac{a}{c} \right)^2 + \frac{1}{c^2} (\pm \rho_f - \pm \rho_s)^2 = 1$$

$$\left(\frac{a}{c} \right)^2 = 1 - \frac{1}{c^2} (\pm \rho_f - \pm \rho_s)^2. \tag{2.68}$$

This can be used to test for a feasible solution, by

$$F = 1 - \frac{1}{c^2} (\pm \rho_f - \pm \rho_s)^2$$

$$> 0. \tag{2.69}$$

In order to compute the rotation angle ϕ_s, the equation can be written in the form

$$t_{ct} = \mathbf{R}(\phi_s)' \frac{1}{c} \begin{pmatrix} \beta \\ \gamma \end{pmatrix},$$

$$\mathbf{R}(\phi_s) = \begin{pmatrix} \cos(\phi_s) & -\sin(\phi_s) \\ \sin(\phi_s) & \cos(\phi_s) \end{pmatrix}, \tag{2.70}$$

where

$$\beta = \sqrt{c^2 - (\pm \rho_f - \pm \rho_s)^2},$$

$$\gamma = (\pm \rho_f - \pm \rho_s). \tag{2.71}$$

Expanding this and solving for ϕ_s gives

$$\cos(\phi_s) \frac{\sqrt{c^2 - (\pm \rho_f - \pm \rho_s)^2}}{c} + \sin(\phi_s) \frac{(\pm \rho_f - \pm \rho_s)}{c} = t_{ct1},$$

$$-\sin(\phi_s) \frac{\sqrt{c^2 - (\pm \rho_f - \pm \rho_s)^2}}{c} + \cos(\phi_s) \frac{(\pm \rho_f - \pm \rho_s)}{c} = t_{ct2}, \tag{2.72}$$

or

$$\frac{1}{c} \left(\begin{array}{cc} \sqrt{c^2 - (\pm\rho_f - \pm\rho_s)^2} & (\pm\rho_f - \pm\rho_s) \\ -(\pm\rho_f - \pm\rho_s) & \sqrt{c^2 - (\pm\rho_f - \pm\rho_s)^2} \end{array} \right) \left(\begin{array}{c} \cos(\phi_s) \\ \sin(\phi_s) \end{array} \right) = t_{ct}. \quad (2.73)$$

Solving for ϕ_s gives

$$\left(\begin{array}{c} \cos(\phi_s) \\ \sin(\phi_s) \end{array} \right) = \frac{1}{c} \left(\begin{array}{cc} \beta & -\gamma \\ \gamma & \beta \end{array} \right) t_{ct}.$$

Hence we obtain

$$\phi_s = \tan^{-1}(\sin(\phi_s), \cos(\phi_s)). \quad (2.74)$$

The final angle ϕ_f can then be determined using

$$\phi_t = \phi_s + \phi_f,$$
$$\phi_f = \phi_t - \phi_f. \quad (2.75)$$

An alternative solution is

$$\mathbf{R}(\phi_s)t_{ct} = \frac{1}{c} \left(\begin{array}{c} \beta \\ \gamma \end{array} \right). \quad (2.76)$$

Expanding this gives

$$\left(\begin{array}{c} \cos(\phi_s) \\ \sin(\phi_s) \end{array} \right) = \frac{1}{\Delta} \left(\begin{array}{cc} t_{ct1} & t_{ct2} \\ -t_{ct2} & t_{ct1} \end{array} \right) \frac{1}{c} \left(\begin{array}{c} \beta \\ \gamma \end{array} \right)$$
$$\Delta = t_{ct1}^2 + t_{ct2}^2$$
$$= 1. \quad (2.77)$$

As the solution is not in a closed form, an iterative solution is required. The known variables at the start are given in Table 2.3.

Table 2.3 Initial variables

Variable	Description
ϕ_t	The angles between the start and finish pose
P_s, P_f	The start and finish position
κ_s, κ_f	The start and finish curvature

From these, the start angle is initialised with a positive and negative value as

$$\phi_{s_p} = \tfrac{1}{2}\phi_t + \delta_s,$$

$$\phi_{s_m} = \tfrac{1}{2}\phi_t - \delta_s, \qquad (2.78)$$

$$\delta_s = \tfrac{1}{4}\phi_t.$$

The iterative solution then follows these steps:

1. Calculate the final arc angles ϕ_{f_p} and ϕ_{f_m} using equation (2.75).
2. Calculate ρ_{s_p}, ρ_{s_m}, α_{s_p} and α_{s_m} variables using equation (2.52).
3. Calculate ρ_{f_p}, ρ_{f_m}, α_{f_p} and α_{f_m} variables using equation (2.54).
4. Calculate the length of the centre vector c_p and c_m using equation (2.61).
5. Test for feasible solutions F_p and F_m using equation (2.69) for the positive and negative values.
6. If neither value is feasible, then:
   (a) calculate a new solution ϕ_n using the Euler algorithm, as shown in Figure 2.13;
   (b) if $F_p > F_m$, then $\phi_m = \phi_n$, else $\phi_p = \phi_n$, as also shown in Figure 2.13;
   (c) goto to the beginning and recompute.
7. Choose the feasible solution F that has the greatest positive value.

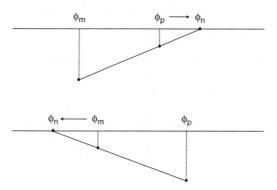

Figure 2.13   Euler interpolation

Note: in order to compute the Fresnel integrals, use series expansions of the form

$$C(s) = \sum_{n=0}^{\infty} \frac{(-1)^n}{(2n+1)!(4n+3)} s^{4n+3},$$

$$S(s) = \sum_{n=0}^{\infty} \frac{(-1)^n}{(2n)!(4n+1)} s^{4n+1}. \tag{2.79}$$

Each element in the series can be added until the solution changes by a prescribed small percentage (for example, 1% change or less).

Once ϕ_s and ϕ_f are found, then the solution is complete. The x_s, y_s position and the x_f, y_f position of the straight-line segment can be determined, together with the arc trajectories. The length of the clothoid path is given by the sum of the clothoid arc segments and the tangent line connecting these arcs.

2.8 Producing Flyable Pythagorean Hodograph Paths (2D)

The two path formats studied, namely Dubins and clothoid, are formed from arc and straight-line segments. Each segment of the path provides either of these basic motions. An alternative path that provides continuous curvature is an alternative solution to the path planning problem. There are a wide variety of polynomials that have the minimum requirement of continuous curvature: that of a non-vanishing second derivative. One of the most used polynomial forms in path planning is that of the B-spline curve.

A B-spline path whose coefficients are calculated using sequential quadratic programming to meet the constraints and objectives is detailed in Lian and Murray (2002). However, curvature constraints are not considered in the paper and, as the curvature of any path must meet the UAV constraints, further analysis is required.

The Pythagorean hodograph (PH) was first introduced by Farouki and Sakkalis (1990). All PH curves are defined by polynomial curves which have hodographs that satisfy a Pythagorean condition. Hodographs are tangent curves that are first derivatives of the curve and hence equate with velocity vectors. The Pythagorean condition states that the sum of the squares of the sides of a right-angled triangle is equal to the square of its hypotenuse. In the time domain, the hodograph is called the velocity vector, which is always parallel to the tangent of the path. However, the derivation of the PH path arose from the definition of path length. The length h of the path $r(q)$

parametrized with parameter q is

$$h(q) = \int_{q_1}^{q_2} \|\dot{r}(q)\| \, dq \tag{2.80a}$$

$$= \int_{q_1}^{q_2} \sqrt{\dot{x}(q)^2 + \dot{y}(q)^2} \, dq, \tag{2.80b}$$

where $q \in [q_1, q_2]$, and $\dot{x}(q) = dx/dq$ and $\dot{y}(q) = dy/dq$ are hodographs.

The calculation of the path length requires the solution to the integral in equation (2.80). The presence of the square root term in the equation may not result in a closed-form solution in general, and the solution must be computed by iteration, with its associated accuracy and computational load problems. A better solution of the path length equation is required to eliminate this problem. Using a special form of polynomial, the Pythagorean hodograph has a closed-form solution. This is obtained by noting that the term inside the square root of equation (2.80) is the sum of the squares of the hodographs. If it is possible to represent the term inside the square root as a perfect square, say $\sigma(q)^2$, then the solution to the path length will be an integral of the polynomial equation $\sigma(t)$, and takes the form

$$\sigma(q)^2 = \dot{x}(q)^2 + \dot{y}(q)^2 \tag{2.81}$$

and

$$h(q) = \int_{q_1}^{q_2} |\sigma(q)| \, dq. \tag{2.82}$$

This is equivalent to satisfying Pythagoras's theorem by taking polynomials $\sigma(q)$ as the hypotenuse, and with $\dot{x}(q)$ and $\dot{y}(q)$ as the two other sides. A polynomial curve whose hodographs meet the condition of equation (2.81) is called the Pythagorean hodograph (PH). The PH path is designed by selecting suitable polynomials $u(q)$, $v(q)$ and $w(q)$ to construct the hodographs $\dot{x}(q)$ and $\dot{y}(q)$, such that they meet the condition in equation (2.81). The functions that meet the PH requirements are given by

$$\dot{x}(q) = w(q)[u(q)^2 - v(q)^2], \tag{2.83}$$

$$\dot{y}(q) = 2w(q)u(q)v(q), \tag{2.84}$$

which imply that

$$\sqrt{\dot{x}(q)^2 + \dot{y}(q)^2} = w(q)[u(q)^2 + v(q)^2]$$

$$= |\sigma(q)|,$$

where $u(q)$ and $v(q)$ are relatively prime polynomials of degree 2 with $w(q) = 1$. This results in a polynomial $\sigma(q)$ of degree $(n-1)$. Now, the problem is reduced to finding the coefficients of the polynomials $u(q)$, $v(q)$ and $w(q)$. The advantage of this format is not only that it produces a closed form to equation (2.80), but also that it results in an equal distribution of control points on the path. In other words, there is an equal increment of path length for an equal increment of the parameter q.

The parametric speed $\dot{q}$ and the curvature $\kappa(q)$ of the PH curve are rational and are given by

$$\dot{h}(q) = |\sigma(q)|, \tag{2.85}$$

$$\kappa = \frac{2[u(q)\dot{v}(q) - \dot{u}(q)v(q)]}{w(q)[u(q)^2 + v(q)^2]}, \tag{2.86}$$

$$r_d(q) = r(q) \pm d n(q), \tag{2.87}$$

where $\dot{h}(q) = dh/dq$ and $n(q)$ is the unit normal to the curve $r(q)$. As an extra bonus, a curve offset by distance d from the PH path is also a PH path. This will be useful when placing a safety distance around the path to test for collision avoidance.

From equation (2.85), the parametric speed of the PH curve is simply a root-finding problem of a polynomial, with the curvature given by equation (2.86). The offset curve in equation (2.87) can be used to define a safety region or sensor range of uncertainty along the path. Figure 2.14 shows a comparative visualisation of a smooth PH path and a Dubins path with the same maximum bound on curvature. It is evident from this figure that the curvature continuity of the PH path is achieved with increased path length. The length of the PH path must be greater than that of the Dubins path, as the Dubins path is the shortest path between any two poses. The safety distance around the PH path is defined by the two offset paths shown by the dotted circles. The minimum order polynomial that exhibits the PH behaviour is the third, and it is called the cubic PH. However, the lowest order of the PH path that has a point of inflection is the fifth (Farouki and Sakkalis 1990), called the quintic PH. The presence of an inflection point allows the path to have more flexibility so that the path can easily be manipulated. Hence, the quintic PH curve is used for path planning. From now on, a PH path denotes a quintic PH curve.

The initial and final positions respectively are (x_s, y_s) and (x_f, y_f), and the corresponding orientations (tangential directions) are ϕ_s and ϕ_f. These variables define the start and finish poses and are boundary values for the PH path.

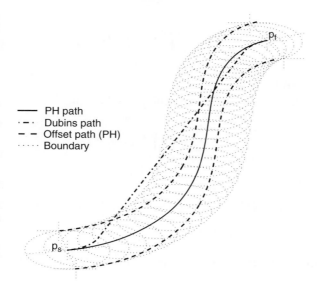

Figure 2.14 Comparison of a Dubins path with a Pythagorean hodograph path. The Dubins path (— · —) is the shortest path but it lacks the curvature continuity. The PH path (——) has continuity but is longer for the same curvature bound. Reproduced by permission of ASME

The PH path is represented in Bézier form for numerical stability. The general equation for an nth-order polynomial in Bézier form is

$$r(q) = \sum_{k=0}^{n} b_k \binom{n}{k} q^k (1-q)^{(n-k)}, \qquad q \in [0, 1], \tag{2.88}$$

where $b_k = (x_k, y_k)$, $k = 1, \ldots, n$, are control points,

$$\binom{n}{k} = \frac{n!}{k!(n-k)!}$$

and $r(q)|_{q=0}$ and $r(q)|_{q=1}$, respectively, represent the starting and ending points of the path. Note that the path parameter q is normalised to unity.

Now, the rth derivative of the nth-order Bézier curve is

$$\frac{d^r r(q)}{dq^r} = \frac{n!}{(n-r)!} \sum_{j=0}^{n-r} \Delta^r b_j \binom{n-r}{j} q^j (1-q)^{(n-r-j)}, \tag{2.89}$$

where

$$\Delta^r b_i = \sum_{j=0}^{r} \binom{r}{j}(-1)^{r-j} b_{i+j},$$

$\Delta b_j = b_{j+1} - b_j$ and $\Delta^0 b_j = b_j$.

For a quintic path, $n = 5$ and equation (2.88) becomes

$$r(q) = \sum_{k=0}^{5} b_k \binom{5}{k} q^k (1-q)^{(5-k)} \tag{2.90}$$

or

$$r(q) = b_0(1-q)^5 + 5b_1 q(1-q)^4 + 10b_2 q^2(1-q)^3$$
$$+ 10b_3 q^3(1-q)^2 + 5b_4 q^4(1-q) + b_5 q^5. \tag{2.91}$$

From equation (2.89), the first derivative of the path $r(q)$ is

$$\frac{dr(q)}{dq} = 5\sum_{j=0}^{4} \binom{4}{j} \Delta^1 b_j q^j (1-q)^{4-j} \tag{2.92}$$

or

$$\frac{dr(q)}{dq} = 5(b_1 - b_0)(1-q)^4 + 20(b_2 - b_1)q(1-q)^3$$
$$+ 30(b_4 - b_3)q^3(1-q) + 5(b_5 - b_4)q^4. \tag{2.93}$$

Now these equations can be used to interpolate between the two poses and to define the shape of the connecting curve. As the position and direction at initial and final locations are known, we use first-order Hermite interpolation. Substituting the position coordinates at $t = 0$ and at $t = 1$ in equation (2.91), and using the first derivative of the path in equation (2.93), the control points b_0, b_1, b_5 and b_4 are calculated as

$$b_0 = (x_s, y_s), \tag{2.94a}$$

$$b_5 = (x_f, y_f), \tag{2.94b}$$

$$d_0 = (\cos(\phi_s), \sin(\phi_s)), \tag{2.94c}$$

$$d_5 = (\cos(\phi_f), \sin(\phi_f)), \tag{2.94d}$$

$$b_1 = b_0 + \tfrac{1}{5}d_0, \tag{2.94e}$$

$$b_4 = b_5 - \tfrac{1}{5}d_5. \tag{2.94f}$$

Thus, the control points (b_0, b_1, b_4, b_5) in equation (2.94) are fixed by the start and finish poses. Now the problem is reduced to finding the two control points b_2 and b_3 so that equation (2.91) satisfies the PH condition in equation (2.81). This results in four solutions (Farouki and Neff 1995). A minimum-energy curve solution from Farouki (1996), which has smooth variation of curvature, is used for path planning.

2.8.1 Design of Flyable Path using 2D PH curve

The PH path obtained from the above equations provides only the tangent continuity. In other words, the development of the path is based on matching the tangents at the end points. For a flexible algorithm to produce a flyable path, tangent continuity is insufficient. Hence, some further refinement is required. One approach found in Bruyninckx and Reynaerts (1997) achieves curvature continuity and improves the range of paths by increasing the length of the boundary vectors. The lengths of the tangent vectors are increased by approximating the term $\partial\kappa/\partial c_t$, where c_t is the magnitude of the tangent vector. However, there is no closed-form solution. Here, the lengths of the tangent vectors are increased directly by modifying equations (2.94c) and (2.94d) to give

$$d_0 = c_0(\cos(\phi_s), \sin(\phi_s)), \tag{2.95}$$

$$d_5 = c_5(\cos(\phi_f), \sin(\phi_f)), \tag{2.96}$$

where $c_0 \in [1, \infty]$ and $c_5 \in [1, \infty]$.

Increasing the values of c_0 and c_5 will increase the length of the tangent vectors $|\overrightarrow{b_0 b_1}|$ and $|\overrightarrow{b_5 b_4}|$ and in turn b_2 and b_3 are modified to meet the PH condition. As there is no closed-form solution available for the modified tangent solution, an iterative method is required. This is achieved by matching a flyable PH path to that of the Dubins path, so that the curvature constraint is met and a flyable path is produced. Figure 2.15 shows the flexibility and evolution of the resulting PH paths to match the Dubins path. The initial PH path has only tangent continuity. The control points P_{21} and P_{31} are then shifted, respectively, to P_{22} and P_{32} by increasing the lengths of the boundary tangent vectors $\overrightarrow{P_s P_{21}}$ and $\overrightarrow{P_f P_{31}}$. This results in a flyable path that meets the maximum-curvature bound.

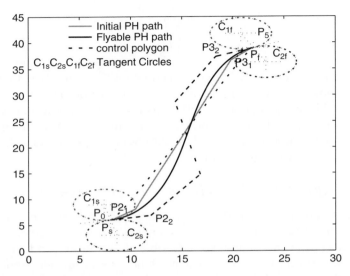

Figure 2.15 Evolution of a PH path from the tangent continuity into curvature continuity. Reproduced from IEEE

References

Boissonnat, J. D., Cerezo, A. and Leblond, J. 1994. Shortest paths of bounded curvature in the plane. *International Journal of Intelligent Systems*, **10**, 1–16.

Bruyninckx, H. and Reynaerts, D. 1997. Path planning for mobile and hyper-redundant robots using pythagorean hodograph curves. *Proc. 8th Int. Conf. on Advanced Robotics, ICAR'97*, pp. 595–600.

Dubins, L. E. 1957. On curves of minimal length with a constraint on average curvature and with prescribed initial and terminal positions and tangent. *American Journal of Mathematics*, **79**, 497–516.

Farouki, R. T. 1996. The elastic bending energy of pythagorean-hodograph curves. *Computer Aided Geometric Design*, **13**, 227–241.

Farouki, R. T. and Neff, C. A. 1995. Hermite interpolation by pythagorean hodograph quintics. *Mathematics of Computation*, **64**(212), 1589–1609.

Farouki, R. T. and Sakkalis, T. 1990. Phythagorean hodographs. *IBM Journal of Research and Development*, **34**(5), 736–752.

Fredman, M. L. and Tarjan, R. E. 1987. Fibonacci heaps and their uses in improved network optimisation algorithms. *Journal of the ACM*, **34**(3), 596–615.

Lian, F.-L. and Murray, R. 2002. Real-time trajectory generation for the cooperative path planning of multi-vehicle systems. *Proc. 41st IEEE Conf. on Decision and Control*, vol. 4, pp. 3766–3769.

Reeds, J. A. and Shepp, R. A. 1990. Optimal paths for a car that goes both forward and backward. *Pacific Journal of Mathematics*, **145**(2), 367–393.

Shkel, A. M. and Lumelsky, V. 2001. Classification of the Dubins set. *Robotics and Autonomous Systems*, **34**, 179–202.

Sussmann, H. and Tang, G. 1991. Shortest paths for the Reeds–Shepp car: a worked out example of the use of geometric techniques in nonlinear optimal control. Technical Report, Department of Mathematics, Rutgers University.

Wong, H., Kapila, V. and Vaidyanathan, R. 2004. UAV optimal path planning using C–C–C class paths for target touring. *Proc. 43rd IEEE Conf. on Decision and Control*, Bahamas, pp. 1105–1110.

3

Path Planning in Three Dimensions

The previous chapter developed the theory and application of path planning in two dimensions. In practice, manoeuvres take place in three dimensions. A three-dimensional manoeuvre will be required, for example, when UAVs take off or land, when they are changing altitude or when they need to fly over, rather than around, obstacles.

Path planning using two-dimensional space is abundant in the literature, which may be due to the fact that path planning is widely studied in ground robotics. However, when dealing with aerial vehicles, altitude has to be added to the planar movement for manoeuvres in space. There are only a few references to work done on path planning in three dimensions. An extension of sample based motion planning into higher-dimensional space is studied in Hsu *et al.* (1997) and Kavraki *et al.* (1996). A direct extension of the Dubins path into three-dimensional manoeuvres is studied in Shanmugavel *et al.* (2006). A similar approach can be seen in Pachikara *et al.* (2009).

From the perspective of path planning, the input to the path planner remains the same, that is, a start and a finish pose. The poses now have three position coordinates, (x, y, z), and three angular orientations, (roll, pitch, yaw). This chapter details algorithms for the production of 3D flight paths. Two techniques will de examined, similar to those of the 2D algorithms in Chapter 2, namely (i) Dubins paths and (ii) spatial Pythagorean

Cooperative Path Planning of Unmanned Aerial Vehicles
Antonios Tsourdos, Brian White and Madhavan Shanmugavel
© 2011 John Wiley & Sons, Ltd

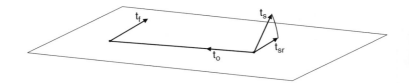

Figure 3.1 Three-dimensional Dubins manoeuvre conditions

hodographs. A three-dimensional flyable path $r(q)$ can be produced by solving the following equation:

$$P_s(x_s, y_s, z_s, \theta_s, \psi_s) \xrightarrow{r(q)} P_f(x_f, y_f, z_f, \theta_f, \psi_f), \qquad (3.1)$$

$$|\kappa(t)| < \kappa_{max}, \qquad |\tau(t)| < \tau_{max}.$$

Here $\kappa(t)$ is the curvature and $\tau(t)$ is the torsion of the path (see Chapter 1).

One of the classical paths used for an aircraft manoeuvre from one altitude to another is a section of the circular helix, whose projection on the $X-Y$ plane is a circle. The path can be visualised as being wound on to the surface of a vertical cylinder. An important property of this curve is that both the curvature and the torsion are constant, with the ratio of the two giving the radius of the cylinder and the pitch of the helix. Compared with the Dubins and PH paths, the path length of the helix will be greater than either and is more restrictive in path shape.

A more useful curve set is to generalise the Dubins and Pythagorean hodograph paths from 2D to 3D. If the two tangent vectors t_s and t_f are joined by a line vector t_o, as shown in Figure 3.1, then a 2D manoeuvre can be performed provided that all of the vectors are coplanar. The figure shows that the two tangent vectors are not coplanar, as the start tangent vector t_s does not lie in the plane defined by the two vectors t_f and t_o.

One approach to producing a 3D path is first to define an initial arc manoeuvre that takes the tangent vector t_s into the plane to get t_{sr} (Figure 3.2). This will then ensure that this vector is coplanar with t_o and t_f and a 2D Dubins path can be computed using the rotated vector t_{sr} as the new start tangent vector. The normal vector n_{sr} can be defined as normal to the tangent vector and also lying in the plane defined by t_f and t_o. The binormal vector b_{sr} then makes up a right-handed set. The initial manoeuvre will then consist of a combined curvature manoeuvre about the binormal vector, followed by a roll manoeuvre to line up the normal and binormal vectors to start the final 2D Dubins manoeuvre. Hence the composite Dubins manoeuvre

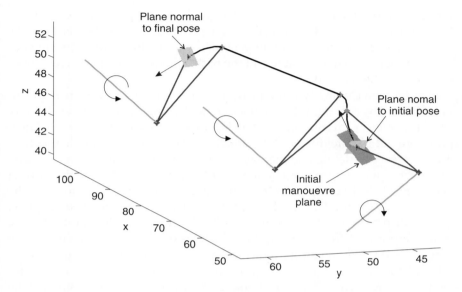

Figure 3.2 Three-dimensional Dubins manoeuvre of a UAV

consists of a coplanar circle manoeuvre with defined curvature κ_s, followed by a normal start manoeuvre circle with defined curvature κ_s, a straight-line manoeuvre and finally a finish manoeuvre circle with defined curvature κ_f. The geometry that defines the final 2D manoeuvre is given in the previous chapter in section 2.5. Figure 3.2 shows the resulting 3D Dubins path.

A more direct approach is to eliminate the initial rotation into a coplanar configuration by performing a full 3D manoeuvre using 3D geometric analysis. Two approaches will be detailed in this chapter. The first is a true 3D Dubins solution using differential geometric concepts, and the other is to use a 3D spatial PH path, which is obtained in a similar manner to the 2D algorithm by use of first-order Hermite interpolation. The resulting path is further tuned to make the paths flyable by increasing the lengths of the boundary tangent vectors. Both of these approaches are detailed in this chapter.

3.1 Dubins Paths in Three Dimensions Using Differential Geometry

As already stated, in the case of 2D Dubins path, the frame (t_s, n_s) and (t_f, n_f) lie in the same manoeuvre plane. That is, the binormal vectors of both the poses are normal to the manoeuvre plane and are thus parallel to each other. But, for many poses in 3D space, this is not generally the case (see Figure 3.1). The two tangent vectors do not in general lie in a common plane, and hence a 2D manoeuvre is not possible and the path generation in 3D space is not as simple as is the case with 2D.

In order for a Dubins-type manoeuvre to take place, that is, an arc manoeuvre followed by a straight-line manoeuvre with a final arc manoeuvre, a more complex analysis is required. As the binormal vectors for the start and finish poses are not parallel for the 3D manoeuvre, for a Dubins-type manoeuvre, it is necessary to define two manoeuvre planes to define the 3D manoeuvre. The start manoeuvre plane contains the tangent vector t_s and the normal vector n_s. This is completed by defining a right-handed set to give the binormal vector b_s. This triple is $[\,t_s n_s b_s\,]$. The second manoeuvre plane is the finish manoeuvre plane defined by the triple $[\,t_f n_f b_f\,]$. Both of these frames are Frenet frames using the principles of differential geometry (see Appendix A). The intersection of the start and finish manoeuvre planes is a line, hence the straight-line manoeuvre is uniquely defined for this case along this intersection. The line intersection is a function of the position of the manoeuvre planes, which are not in general defined, as the UAV can perform an initial roll manoeuvre to establish the start manoeuvre plane and a final roll manoeuvre to align the final tangent, normal and binormal vector set.

The sign of the manoeuvre can be determined by considering the centreline between the two positions. Viewed from each position, a positive or negative rotation from the tangent vector to the centre vector will define the sign of the curvature for each manoeuvre. The start manoeuvre and the finish manoeuvre planes are obtained by rotation about the tangent vector t_s. So, we have

$$
\begin{aligned}
\left[\begin{array}{ccc} t_{ms} & n_{ms} & b_{ms} \end{array}\right] &= \left[\begin{array}{ccc} t_s & n_s & b_s \end{array}\right] \mathbf{R}_s, \\
\left[\begin{array}{ccc} t_{mf} & n_{mf} & b_{mf} \end{array}\right] &= \left[\begin{array}{ccc} t_f & n_f & b_f \end{array}\right] \mathbf{R}_f,
\end{aligned}
\tag{3.2}
$$

where

$$
\mathbf{R}_s = \begin{pmatrix} 1 & 0 & 0 \\ 0 & \cos(\phi_s) & -\sin(\phi_s) \\ 0 & \sin(\phi_s) & \cos(\phi_s) \end{pmatrix},
$$

$$
\mathbf{R}_f = \begin{pmatrix} 1 & 0 & 0 \\ 0 & \cos(\phi_f) & -\sin(\phi_f) \\ 0 & \sin(\phi_f) & \cos(\phi_f) \end{pmatrix},
\tag{3.3}
$$

and where ϕ_s and ϕ_f are the rotation angles for the start and finish manoeuvre planes.

The radius vectors can then be defined in the manoeuvre planes as

$$r_s = \begin{bmatrix} t_{ms} & n_{ms} & b_{ms} \end{bmatrix} \begin{pmatrix} 0 \\ \pm 1/\kappa_s \\ 0 \end{pmatrix} \tag{3.4}$$

and, similarly,

$$r_f = \begin{bmatrix} t_{mf} & n_{mf} & b_{mf} \end{bmatrix} \begin{pmatrix} 0 \\ \pm 1/\kappa_f \end{pmatrix}. \tag{3.5}$$

The Frenet basis vectors are related by

$$\begin{bmatrix} t_f & n_f & b_f \end{bmatrix} = \begin{bmatrix} t_s & n_s & t_s \end{bmatrix} R, \tag{3.6}$$

where R is the rotation matrix required to change the axis set from start to finish axes.

Hence, we have

$$R = \begin{pmatrix} t_f & n_f & b_f \end{pmatrix} \cdot \begin{pmatrix} t_s & n_s & b_s \end{pmatrix}, \tag{3.7}$$

giving

$$R = \begin{pmatrix} t_f \cdot t_s & t_f \cdot n_s & t_f \cdot b_s \\ n_f \cdot t_s & n_f \cdot n_s & n_f \cdot b_s \\ b_f \cdot t_s & b_f \cdot n_s & b_f \cdot b_s \end{pmatrix}. \tag{3.8}$$

The connecting vectors a_s, a_f and a_c form an orthogonal set of vectors. The connecting vectors a_s and a_f are normal to the vector a_c, but are not parallel. Each vector lies in the appropriate manoeuvre plane, which are not coincident. The internal connecting vector a_c is common to both manoeuvre planes. It can thus be written in the form

$$a_c = a \begin{bmatrix} t_{ms} & n_{ms} & b_{ms} \end{bmatrix} \alpha_s,$$

$$= a \begin{bmatrix} t_{mf} & n_{mf} & b_{mf} \end{bmatrix} \alpha_f, \tag{3.9}$$

with

$$\alpha_s = \begin{pmatrix} \alpha_{ts} \\ \alpha_{ns} \\ \alpha_{bs} \end{pmatrix} \quad \text{and} \quad \alpha_f = \begin{pmatrix} \alpha_{tf} \\ \alpha_{nf} \\ \alpha_{bf} \end{pmatrix}.$$

The Frenet frames for the two manoeuvre planes can be related by

$$
\begin{aligned}
\begin{bmatrix} t_f & n_f & b_f \end{bmatrix} &= \begin{bmatrix} t_s & n_s & b_s \end{bmatrix} \mathbf{R}, \\
\begin{bmatrix} t_{mf} & n_{mf} & b_{mf} \end{bmatrix} &= \begin{bmatrix} t_f & n_f & b_f \end{bmatrix} \mathbf{R}_f, \\
\begin{bmatrix} t_{ms} & n_{ms} & b_{ms} \end{bmatrix} &= \begin{bmatrix} t_s & n_s & b_s \end{bmatrix} \mathbf{R}_s.
\end{aligned}
\tag{3.10}
$$

Hence we have

$$
\begin{aligned}
\begin{bmatrix} t_{ms} & n_{ms} & b_{ms} \end{bmatrix} \mathbf{R}_s' &= \begin{bmatrix} t_s & n_s & b_s \end{bmatrix}, \\
\begin{bmatrix} t_{mf} & n_{mf} & b_{mf} \end{bmatrix} \mathbf{R}_f' &= \begin{bmatrix} t_f & n_f & b_f \end{bmatrix} \\
&= \begin{bmatrix} t_s & n_s & b_s \end{bmatrix} \mathbf{R} \\
&= \begin{bmatrix} t_{ms} & n_{ms} & b_{ms} \end{bmatrix} \mathbf{R}_s' \mathbf{R},
\end{aligned}
\tag{3.11}
$$

and so

$$
\begin{bmatrix} t_{mf} & n_{mf} & b_{mf} \end{bmatrix} = \begin{bmatrix} t_{ms} & n_{ms} & b_{ms} \end{bmatrix} \mathbf{R}_s' \mathbf{R} \mathbf{R}_f.
\tag{3.12}
$$

This implies that

$$
\begin{aligned}
\alpha_s &= \mathbf{R}_s' \mathbf{R} \mathbf{R}_f \alpha_f, \\
\alpha_f &= \mathbf{R}_f' \mathbf{R} \mathbf{R}_s \alpha_s.
\end{aligned}
\tag{3.13}
$$

The radius vectors r_s and r_f can also be described in start manoeuvre axes, to give

$$
r_s = \begin{bmatrix} t_{ms} & n_{ms} & b_{ms} \end{bmatrix} \begin{pmatrix} 0 \\ \pm 1/\kappa_s \\ 0 \end{pmatrix},
$$

$$
r_f = \begin{bmatrix} t_{ms} & n_{ms} & b_{ms} \end{bmatrix} \mathbf{R}_s' \mathbf{R} \mathbf{R}_f \begin{pmatrix} 0 \\ \pm 1/\kappa_f \\ 0 \end{pmatrix}.
\tag{3.14}
$$

Now, the vectors a_s and a_f lie in the manoeuvre planes and are normal to the connecting vector a_c. These can also be defined in start manoeuvre axes, in the form

$$
a_s = \frac{\pm 1}{\kappa_s} \begin{bmatrix} t_{ms} & n_{ms} & b_{ms} \end{bmatrix} \beta_s,
$$

$$a_f = \frac{\pm 1}{\kappa_f} \begin{bmatrix} t_{mf} & n_{mf} & b_{mf} \end{bmatrix} \beta_f, \tag{3.15}$$

$$= \frac{\pm 1}{\kappa_f} \begin{bmatrix} t_{ms} & n_{ms} & b_{ms} \end{bmatrix} \mathbf{R}'_s \mathbf{R} \mathbf{R}_f \beta_s,$$

where, to ensure that the connection vectors lie in the manoeuvre plane, and are normal to the internal connection vector a_c, we have

$$\beta_s = \frac{1}{b_s} \begin{pmatrix} -\alpha_{ns} \\ \alpha_{ts} \\ 0 \end{pmatrix}, \qquad \beta_f = \frac{1}{b_f} \begin{pmatrix} -\alpha_{nf} \\ \alpha_{tf} \\ 0 \end{pmatrix},$$

$$b_s = \sqrt{\alpha_{ns}^2 + \alpha_{ts}^2}, \qquad b_f = \sqrt{\alpha_{nf}^2 + \alpha_{tf}^2}, \tag{3.16}$$

$$\beta_s \alpha_s = 0, \qquad \beta_f \alpha_f = 0.$$

The position of the finish point p_f relative to the start position p_s is measured in start plane axes $[\, t_s n_s b_s \,]$, so that

$$p_f - p_s = \begin{bmatrix} t_s & n_s & b_s \end{bmatrix} p$$

$$= \begin{bmatrix} t_{ms} & n_{ms} & b_{ms} \end{bmatrix} \mathbf{R}'_s p,$$

$$p_m = \mathbf{R}'_s p, \tag{3.17}$$

$$p = \begin{pmatrix} p_t \\ p_n \\ p_b \end{pmatrix}.$$

Then, the vector sum for the position vector is given by

$$p_m = -r_s + a_s + a_c - a_f + r_f$$

$$p_m + r_s - r_f = a_s + a_c - a_f. \tag{3.18}$$

Substituting for p_m and r_f gives

$$\mathbf{R}'_s p + r_s - \mathbf{R}'_s \mathbf{R} \mathbf{R}'_f r_f = a_s + a_c - \mathbf{R}'_s \mathbf{R} \mathbf{R}_f a_f. \tag{3.19}$$

Rewriting this in the start plane axes gives

$$p + \mathbf{R}_s r_s - \mathbf{R} \mathbf{R}'_f r_f = \mathbf{R}_s a_s + \mathbf{R}_s a_c - \mathbf{R} \mathbf{R}_f a_f. \tag{3.20}$$

3.2 Path Length–Dubins 3D

The Dubins path in 3D forms a composite path of either three or four segments. The path length of the Dubins path, h_{Dubins}, is the sum of the length of these segments and is given by

$$h_{Dubins} = h_i + h_s + a_t + h_f$$

$$= \frac{\alpha_i}{\kappa_s} + \frac{\alpha_s}{\kappa_s} + a + \frac{\alpha_f}{\kappa_f}, \qquad (3.21)$$

where h represents the length of the path, and the suffixes i, s, t and f, respectively, are the initial arc, the start arc, the straight line of length a and the final arc segments. Also α and κ are the arc angles and their associated curvatures. For the Dubins path with the initial manoeuvre into a coplanar solution, the first term h_i is included. For the more direct differential geometric solution, this term is omitted.

3.3 Pythagorean Hodograph Paths–3D

A Pythagorean hodograph is a polynomial curve first introduced by Farouki (Farouki and Sakkalis 1990, 1994; Farouki et al. 2002). A comprehensive treatment of the 2D case was dealt with in Chapter 2 in section 2.8. This chapter will extend the treatment to 3D. As with the 2D case, a fifth-order PH curve is used, as this is the lowest-order curve that has inflection points that provide sufficient flexibility in path shape to be appropriate for path planning (Farouki and Sakkalis 1990). As seen in section 2.8, the PH path provides closed-form polynomials for path length, its curvature and the offset curve. Substituting an appropriate polynomial $\sigma(t)$ such that the variable $\sigma(q)^2 = \dot{x}(q)^2 + \dot{y}(q)^2 + \dot{z}(q)^2$ in equation (3.22) produces a path length $h(q)$ and speed $\dot{h}(q)$, which are reduced to an integral of the polynomial $\sigma(q)$ and the polynomial itself, respectively. The canal tube around the path is also rational, which is used to define the safety region around each UAV. The basics of the PH curve are given in Appendix B.

The path length $h(q)$ of the curve $r(q) = \{x(q), y(q), z(q)\}$ is given by

$$h(q) = \int_{q_1}^{q_2} \sqrt{\dot{x}(q)^2 + \dot{y}(q)^2 + \dot{z}(q)^2}\, dq, \qquad q \in [q_1, q_2]. \qquad (3.22)$$

The term inside the square root in equation (3.22) is the sum of the squares of the hodographs. If $x(q)$, $y(q)$ and $z(q)$ are polynomial functions of the path variable q, and we could make this term a perfect square, then the path length would simply be an integral of a polynomial $\sigma(q)$, in a similar manner to the 2D treatment. So

$$\sigma(q)^2 = \dot{x}(q)^2 + \dot{y}(q)^2 + \dot{z}(q)^2 \tag{3.23}$$

and

$$h(q) = \int_{q_1}^{q_2} |\sigma(q)| \, dq. \tag{3.24}$$

For any polynomial curve, if its hodographs meet the Pythagorean condition, the curve is called a Pythagorean hodograph curve. Now, for a polynomial curve, that is, when $x(q)$, $y(q)$ and $z(q)$ are polynomials, a useful PH path is designed by selecting suitable polynomials for the hodographs, $\dot{x}(t)$, $\dot{y}(t)$ and $\dot{z}(t)$. The main advantages of this formulation are: (i) calculation of path length without any approximation; (ii) equal increment of distance travelled along the curve for equal increment of the parameter q: (iii) rational parametric speed; and (iv) rational intrinsic properties (curvature, torsion and canal surface).

3.3.1 Spatial PH Curves

Consider a polynomial space curve $r(q) = (x(q), y(q), z(q))$ represented in pure quaternion form: $r(q) = (x(q)i + y(q)j + z(q)k)$. The curve $r(q)$ is a PH curve only if

$$\frac{dr}{dq} = Q(q)iQ^*(q) \tag{3.25}$$

for some quaternion polynomial $Q(q) = a(q) + ib(q) + jc(q) + kd(q)$. The hodographs of $r(q)$ satisfy

$$\dot{x}(q) = a(q)^2 + b(q)^2 - c(q)^2 - d(q)^2, \tag{3.26a}$$

$$\dot{y}(q) = 2[a(q)d(q) + b(q)c(q)], \tag{3.26b}$$

$$\dot{z}(q) = 2[b(q)d(q) - a(q)c(q)], \tag{3.26c}$$

$$\sigma(q) = a(q)^2 + b(q)^2 + c(q)^2 + d(q)^2. \tag{3.26d}$$

These equations provide necessary and sufficient conditions for a polynomial space curve to be PH. The quaternion $Q(q)$ in Bézier form is

$$Q(q) = \sum_{i=0}^{2} Q_i \binom{2}{i} q^i (1-q)^{2-i}, \qquad q \in [0,1]. \tag{3.27}$$

The coefficients Q_0, Q_1 and Q_2 are found by Hermite interpolation (see Farouki and Sakkalis 1994; Farouki *et al.* 2002). The length of the curve $h(q)$ is

$$h(q) = \int_{q_1}^{q_2} |Q(q)|^2 \, dq. \tag{3.28}$$

A PH curve designed by using equation (3.27) is obtained by interpolating over positions and directions. This is not a smooth path and hence the path has to be smoothed for curvature continuity.

3.4 Design of Flyable Paths Using PH Curves

Equation (3.27) is a quintic polynomial designed by interpolating the free vectors at the boundaries. The free vectors have positions (x, y, z) and direction (θ, ψ) in space. A curve interpolating two such vectors is called a Hermite interpolation. The resulting curve will have tangent continuity but not curvature continuity. For UAV application, it is essential to have curvature continuity, as the curvature is proportional to the lateral acceleration of the UAV. Hence we need to have a controlled curvature at the boundaries of the interpolation curves, as well as to impose a maximum-curvature constraint. This will ensure a flyable path for the UAV.

Now, the PH curve is represented in fifth-order Bernstein–Bézier polynomial form as

$$r(q) = \sum_{k=0}^{5} b_k \binom{5}{k} (1-q)^{(5-k)} q^k, \qquad q \in [0,1], \tag{3.29}$$

where $b_k = (x_k, y_k, z_k)$ are control points, whose vertices define the control polygon or Bézier polygon with $k = 0, \ldots, 5$. The initial and final configurations are $P_s(x_s, y_s, z_s, \theta_s, \psi_s)$ and $P_f(x_f, y_f, z_f, \theta_f, \psi_f)$, respectively. The four control points of the Bézier polygons are calculated by first-order Hermite interpolation as follows:

$$b_0 = (x_s, y_s, z_s), \tag{3.30a}$$

$$b_5 = (x_f, y_f, z_f), \tag{3.30b}$$

$$d_0 = m_0[\cos(\theta_s)\cos(\psi_s), \cos(\theta_s)\sin(\psi_s), \sin(\theta_s)], \tag{3.30c}$$

$$d_5 = m_5[\cos(\theta_f)\cos(\psi_f), \cos(\theta_f)\sin(\psi_f), \sin(\theta_f)], \tag{3.30d}$$

$$b_1 = b_0 + \tfrac{1}{5}d_0, \tag{3.30e}$$

$$b_4 = b_5 - \tfrac{1}{5}d_5, \tag{3.30f}$$

where (x_s, y_s, z_s) is the initial position, (x_f, y_f, z_f) is the final position, (θ_s, ψ_s) is the initial orientation and (θ_f, ψ_f) is the final orientation. The orientations use spherical coordinates. The positive constants m_0 and m_5 play a crucial role in defining the interpolation curve. The constants increase the length of the control vectors $\overrightarrow{b_0 b_1}$ and $\overrightarrow{b_4 b_5}$, which in turn fix the control points b_2 and b_3 satisfying the PH condition (3.26). This changes the curvature and torsion of the path, with a corresponding change in shape. From equation (3.30), the control points (b_0, b_1, b_4, b_5) are fixed. Now the problem is reduced to finding the control points b_2 and b_3. These are found by using equation (3.27).

3.4.1 Design of Flyable Paths

The resulting path is tangent-continuous but not curvature-continuous. In addition to the positions and directions, we also need to interpolate for curvature at the end points. As no closed-form solution exists for curvature interpolation, an iterative process is adopted in order to meet the curvature constraint. The path also needs to be interpolated in order to produce a curve within the maximum curvature bounds over its full length to make it flyable. The rth derivative of the path is

$$\frac{d^r r(q)}{dq^r} = \frac{5!}{(5-r)!} \sum_{j=0}^{5-r} \Delta^r B_j \binom{5-r}{j} q^j (1-q)^{5-r-j}, \qquad j = 0, 1, \ldots, 5, \tag{3.31}$$

where

$$\Delta^r b_i = \sum_{j=0}^{r} \binom{r}{j}(-1)^{r-j} b_{i+j},$$

$$\Delta^0 B_j = B_j, \tag{3.32}$$

$$\Delta^r B_j = \Delta^{k-1} B_{j+1} - \Delta^{k-1} B_j, \qquad k = 1, 2, \ldots, (5-r).$$

Using equation (3.31) the values of the derivatives of the curve at the boundary points are

$$\dot{r}(q)_{q=0} = 5B_{01}, \tag{3.33a}$$

$$\dot{r}(q)_{q=1} = 5B_{54}, \tag{3.33b}$$

$$\ddot{r}(q)_{q=0} = 20(B_{12} + B_{01}), \tag{3.33c}$$

$$\ddot{r}(q)_{q=1} = 20(B_{54} + B34), \tag{3.33d}$$

$$\dddot{r}(q)_{q=0} = 120(B_{03} - 3B_{12}), \tag{3.33e}$$

$$\dddot{r}(q)_{q=1} = 60(B_{52} - 3B_{43}), \tag{3.33f}$$

where $B_{ij} = B_j - B_i$, and

$$\left| \dot{r}(q) \quad \ddot{r}(q) \quad \dddot{r}(q) \right|_{q=0} = \left| 5B_{01} \quad 20(B_{12} + B_{01}) \quad 120(B_{03} - 3B_{12}) \right|, \tag{3.34}$$

$$\left| \dot{r}(q) \quad \ddot{r}(q) \quad \dddot{r}(q) \right|_{q=1} = \left| 5B_{54} \quad 20(B_{54} + B_{34}) \quad 60(B_{52} - 3B_{43}) \right|, \tag{3.35}$$

where $| \ldots |$ represents the determinant.

Using equations (3.33) in equation (A.17) in Appendix A gives

$$\left| \dot{r}(q) \times \ddot{r}(q) \right|_{q=0} = \left| 5B_{01} \times 20B_{12} \right|$$

$$= 100 \left| (B_1 - B_0) \times (B_2 - B_1) \right| \tag{3.36}$$

$$= 100A_0, \tag{3.37}$$

$$\dot{r}(q)_{q=0} = 5(B_1 - B_0), \tag{3.38}$$

with

$$\left| \kappa(q) \right|_{q=0} = \frac{4}{5} \frac{A_0}{\| B_{01} \|^3}, \tag{3.39}$$

where A_0 is the area of the triangle formed by the control points B_0, B_1 and B_2. Similarly, the curvature at the end point that is $q = 1$ is given by

$$\left| \dot{r}(q) \times \ddot{r}(q) \right|_{q=1} = \left| 5B_{34} \times 20B_{45} \right|$$

$$= 100 \left| (B_4 - B_3) \times (B_5 - B_4) \right| \tag{3.40}$$

$$= 100A_1, \tag{3.41}$$

$$\dot{r}(q)_{q=1} = 5(B_5 - B_4), \tag{3.42}$$

with

$$\left| \kappa(t) \right|_{t=1} = \frac{4}{5} \frac{A_1}{\|B_{45}\|^3}, \tag{3.43}$$

where A_1 is the area of the triangle formed by the control points B_5, B_4 and B_3. For maximum curvature κ_{max}, the boundary curvature has to satisfy

$$\kappa_{max} \le \frac{4}{5} \frac{A_0}{\|B_{01}\|^3}, \tag{3.44a}$$

$$\kappa_{max} \le \frac{4}{5} \frac{A_1}{\|B_{45}\|^3}. \tag{3.44b}$$

Using equations (3.33) and (3.35) in equation (A.18), the value of the torsion at the boundary points becomes

$$|\tau(q)_{q=0}| = \frac{\left|5B_{01} \quad 20(B_{12} + B_{01}) \quad 120(B_{03} - 3B_{12})\right|}{|5B_{01} \times 20B_{12}|^2}, \tag{3.45}$$

$$|\tau(q)_{q=1}| = \frac{\left|5B_{54} \quad 20(B_{54} + B_{34}) \quad 60(B_{52} - 3B_{43})\right|}{|5B_{34} \times 20B_{45}|^2}. \tag{3.46}$$

Substituting equations (3.37) and (3.41) into equation (3.46), the maximum torsion τ_{max} at the boundary points has to satisfy

$$|\tau(q)_{q=0}| \le \frac{\left|5B_{01} \quad 20(B_{12} + B_{01}) \quad 120(B_{03} - 3B_{12})\right|}{A_0^2}, \tag{3.47a}$$

$$|\tau(q)_{q=1}| \le \frac{\left|5B_{54} \quad 20(B_{54} + B_{34}) \quad 60(B_{52} - 3B_{43})\right|}{A_1^2}, \tag{3.47b}$$

where A_0 is the area of the triangle formed by the control points B_0, B_1 and B_2, and A_1 is the area of the triangle formed by the control points B_5, B_4 and B_3. Figure 3.3 shows a spatial PH path with a tube around it. The path is curvature-continuous and hence is a flyable path.

Thus a flyable path is designed by interpolating the positions, directions and curvature at the end points. The PH path is thus optimised for the maximum-curvature bound given in equation (3.44). A similar procedure is adopted for the torque optimisation. However, the resulting form (3.47)

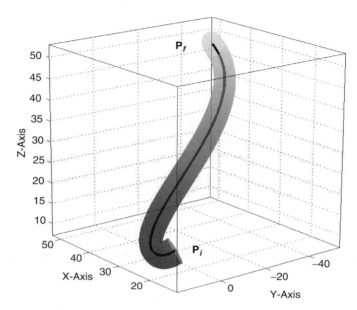

Figure 3.3 Spatial PH path with tube. Reprinted with permission of the American Institute of Aeronautics and Astronautics

cannot be interpreted in such a simple geometrical form as in equation (3.44). Hence an iterative procedure is adopted to arrive at a satisfactory value of the torque. In both cases, the lengths of the tangent vectors are increased by increasing the values of m_0 and m_5 until the path meets the maximum bounds on curvature and torque in equations (3.47) and (3.44). Hence

$$d_0 = m_0[\cos(\theta_s)\cos(\psi_s), \cos(\theta_s)\sin(\psi_s), \sin(\theta_s)], \qquad (3.48a)$$

$$d_5 = m_5[\cos(\theta_f)\cos(\psi_f), \cos(\theta_f)\sin(\psi_f), \sin(\theta_f)], \qquad (3.48b)$$

where $m_0 \in [1, \infty]$ and $m_5 \in [1, \infty]$.

References

Dubins, L. E. 1957. On curves of minimal length with a constraint on average curvature and with prescribed initial and terminal positions and tangent. *American Journal of Mathematics*, **79**, 497–516.

Farouki, R. T. and Sakkalis, T. 1990. Pythagorean hodographs. *IBM Journal of Research and Development*, **34**(5), 736–752.

Farouki, R. T. and Sakkalis, T. 1994. Pythagorean hodograph space curves. *Advances in Computational Mathematics*, **2**, 41–66.

Farouki, R. T., Kandari, M. and Sakkalis, T. 2002. Hermite interpolation by rotation-invariant spatial Pythagorean hodograph curves. *Advances in Computational Mathematics*, **17**, 369–383.

Hsu, D., Latombe, J. C. and Motwani, R. 1997. Path planning in expansive configuration spaces. *IEEE Int. Conf. on Robotics and Automation*, Albuquerque, NM, pp. 2719–2726.

Kavraki, L. E., Svestka, P., Latombe, J. C. and Overmars, M. H. 1996. Probabilistic roadmaps for path planning in high-dimensional configuration spaces. *IEEE Transactions on Robotics and Automation*, **12**(4), 566–580.

Pachikara, A. J., Kehoe, J. J. and Lind, R. 2009. A path-parameterization approach using trajectory primitives for 3-dimensional motion planning. *AIAA Guidance, Navigation, and Control Conf.*, Chicago, August. AIAA 2009-5625.

Shanmugavel, M., Tsourdos, A., Żbikowski, R. and White, B. A. 2006. 3D Dubins sets based coordinated path planning for swarm of UAVs. *AIAA Guidance, Navigation, and Control Conf. and Exhibit*, Keystone, CO, 21–24 August. AIAA-2006-6211.

4

Collision Avoidance

The UAV may be assigned either to map or to avoid obstacles. Here we limit the discussion to avoiding obstacles, as mapping is beyond the scope of the present book. An 'obstacle' will be defined as any object in the environment that the UAV has to avoid. These can be moving or fixed. Fixed obstacles could be objects such a buildings, hills, forests, etc. that will intercept the normal flight path of the UAV. They may also be clear areas that the UAV may not fly into. These would include controlled airspace or dangerous areas. Moving obstacles constitute other aircraft or UAVs that occupy the same airspace and whose trajectories will intercept the UAV flight path. These moving obstacles could be part of the UAV group that would be expected to cooperate with the UAV, but they could also be non-cooperating vehicles that are flying through the airspace occupied by the UAV.

Obstacle avoidance will normally be in the form of a software program within the UAV combined with a set of sensors that monitor the UAV's environment for fixed or moving obstacles. If detected, the software will then modify the UAV path if a future collision is detected.

A collision between two objects occurs when they try to occupy the same point at the same time. As the collision needs to be avoided, a prediction mechanism is required to provide enough lead time to take evasive action. As stated, there are two categories of obstacle: fixed and moving. The first case of fixed obstacles deals with path planning of a single UAV in an environment of static obstacles, where most, if not all, of the obstacles' locations and sizes are known in advance. In the second case of moving obstacles, two possibilities arise: (i) negotiating static obstacles with more than one UAV,

Cooperative Path Planning of Unmanned Aerial Vehicles
Antonios Tsourdos, Brian White and Madhavan Shanmugavel
© 2011 John Wiley & Sons, Ltd

where each UAV has to avoid collision with other UAVs (inter-collision avoidance between UAVs); or (ii) negotiating dynamic obstacles with one or more UAVs, where each UAV has to avoid non-cooperating vehicles in addition to inter-collision avoidance.

The case of static known obstacles is normally handled by global path planners such as Voronoi diagrams, cell decomposition, the establishment of potential fields, or global optimization techniques. The approach taken in this book is to build on the path forms already used for path planning, such as Dubins and PH paths, and to adapt them to negotiate obstacles. This problem is relatively easy to handle, as the obstacles are fixed with respect to place and time, whereas in the case of dynamic obstacles, the trajectories need to be estimated prior to avoidance path planning. However, in both cases, the presence of obstacles causes not only re-planning of paths, but also the management of communications, task assignment, resource allocation and other functions, depending on the mission. These issues are important, but are beyond the scope of this book. Suffice it to say that the use of efficient path planning is a core requirement for these issues.

The basic equation of path planning (6.4) for an obstacle-rich environment is now modified to

$$P_s \xrightarrow{r(q)} P_f, \amalg_{\text{safe}}. \tag{4.1}$$

This effectively states that the path from the start pose P_s to the finish pose P_f should be safe and so will avoid all fixed and moving obstacles. It will be assumed that all fixed obstacles have known positions and all mobile obstacles will have known position and velocity (speed and direction).

When encountering an obstacle, the direction of manoeuvre in 2D is restricted to manoeuvring to either the left or the right of the obstacle, as shown in Figure 4.1. However, the number of choices is greater in three dimensions. For example, if we extend the circle into a sphere, the number of choices is infinite in that any path that manoeuvres around the sphere is a valid path.

Apart from the locations and mobility of the obstacles, the other factors that make the problem complex are the size and shape of the obstacles. Research on the avoidance of convex polygonal obstacles is ongoing in the area of computational geometry, and will not be dealt with in this book, although this problem is very similar to the approach taken here. Producing curvature-constrained paths among the polygonal obstacles is NP-hard (non-deterministic polynomial-time hard) (Agarwal *et al.* 2002; Reif and Wang 1998). For the treatment in this book, it will be assumed that the

Figure 4.1   Obstacle avoidance in 2D

obstacles are surrounded by a safety circle or sphere, which determines the safe boundary of the obstacle for path planning purposes.

This chapter deals with collision avoidance of mapped obstacles, where the obstacles are static and known *a priori*. In the first case, that of static obstacles, two main approaches are used to re-plan the path in 2D. The first is again by analytical geometry, where the obstacle is avoided either by increasing the curvature of the path or by creating an intermediate waypoint and creating a path with several segments. The curvature approach is based on the fact that the curvature and the torsion determine the path in space. In the case of 2D Dubins and clothoid composite paths, the start and finish radii are increased to meet the safety conditions using equation (2.13b). A safe 2D PH path is produced by increasing the boundary curvatures: equations (2.95) and (3.48a), and equations (2.96) and (3.48b). The offset PH path is used to represent the safety boundary on either side of the flight path. Differential geometry is also presented to extend the Dubins paths to include the obstacles in the path structure.

The next chapter discusses dynamic obstacle avoidance and resolution conflicts, where differential geometry is used to predict the mobile obstacle trajectory and to develop geometries that produce trajectories to avoid collision.

4.1 Research into Obstacle Avoidance

The obstacle avoidance problem is closely associated with path planning because the presence of obstacles usually results in the re-planning of paths. Hence, a brief review of approaches to obstacle avoidance is useful here.

The configuration space principle illustrated in Lozano-Pérez (1983) for robotics applications reduces the size of the robots to point masses and compensates for this reduction by increasing the size of the obstacles. This approach enables efficient use of algorithms such as Voronoi diagrams (Beard *et al.* 2002; Bortoff 2000; Li *et al.* 2002; McLain 2000), visibility diagrams and probabilistic methods to model the given map of landmarks and obstacle locations into a searchable database (see Chapter 1). The databases are graphs and trees forming a network of nodes and edges isolating the obstacles. A search algorithm is used to connect the start and finish points through the nodes of the network. The A^* algorithm is used in Yang and Zhao (2004) to find the path in the midst of known obstacles and conflicts. However, the resultant path may not be flyable, as the curvature constraint on paths is not taken into account. A hybrid approach in which a route planned for the known obstacles is deformed to handle the unknown obstacles is used in Lamiraux *et al.* (2004). A similar method of deforming a planned path by varying its curvature is seen in Shanmugavel *et al.* (2005). However, a reactive method, responding to sensor feedback, is essential to handle fast-moving dynamic obstacles.

Obstacle avoidance methods have been studied using potential fields, where attractive and repulsive forces, respectively, are generated by the goal and obstacles (Kim and Khosla 1992). The potential field is then used to avoid the obstacles in path planning of multiple UAVs by flying down the valleys of the potential field. The obstacles are assumed to be circular in shape, with a safety circle surrounding them (Eun and Bang 2006).

An analytical and discrete optimization approach is used in Zabarankin *et al.* (2002) for optimal risk path generation in 2D space, with a constraint on path length. Mixed integer linear programming (MILP) is used in Richards and How (2002) for collision avoidance. The resulting path from both of these approaches does not take into account the curvature constraint, and so extra work has to be done to produce a flyable path. Avoidance circles are used for obstacle avoidance of multiple mobile robots in Fujimori *et al.* (2002), where two robots cooperate within a specified area, while others wait outside the area. However, what is feasible for mobile robots, where the robot can stop and perform pivot turns, is not feasible for UAV applications, where the UAV cannot stop and where they have curvature constraints imposed on their paths.

Path planning with obstacle avoidance in 3D is more complicated. The complication arises because there are infinitely more directions for manoeuvres for both the UAV and the obstacle. Also, other constraints, especially generating the shortest path length, is more complicated in 3D space. An octree representation of 3D space has been proposed in Kitamura *et al.* (1995, 1996), where the space is divided into searchable regions or cells and the collision-free path is generated by applying potential fields to each cell of the octree.

The obstacle avoidance problem has been studied in different applications: to minimize the risk of aircraft detection by radars, sensors or surface-to-air missiles (SAMs) (Chan and Foddy 1985; Hebert *et al.* 2001; Vian and More 1989; Zabarankin *et al.* 2002), to minimize the risk of submarine detection by sensors (Washburn 1990), or to maximize the probability of target detection by a searcher (Assaf and Sharlin-Bilitzky 1994; Benkoski *et al.* 1991; Eagle and Yee 1990; Koopman 1980; Stone 1975; Thomas and Eagle 1995; Washburn 1983). Another application area in this research field is the use of UAVs in civilian airspace, where a collision avoidance algorithm equivalent to that available in commercial aircraft is required. The existing system uses the 'Traffic alert and Collision Avoidance System' (TCAS), which issues resolution advisories (RAs) based on the position and altitude data from transponders aboard the aircraft and from the ground sensors, if available. The safety aspects of using UAVs in civilian airspace is discussed in Dowek *et al.* (2001) and Zeitlin and McLaughlin (2007). The generation of cooperative trajectories for air traffic is discussed in Bicchi and Pallottino (2000). This is an active research area and will need to be developed to a point where such algorithms are robust and accepted before UAVs can fly in controlled airspace with civilian aircraft.

4.2 Obstacle Avoidance for Mapped Obstacles

In order to manoeuvre in an environment that is cluttered with obstacles, the UAV needs to be able to compute trajectories that will circumnavigate these obstacles and still arrive at the final pose and position. Hence there is a requirement to compute composite Dubins trajectories that include avoidance manoeuvres.

Consider the simple case of a single obstacle that intersects the straight-line segment of the standard Dubins trajectory, as shown in Figure 4.2. The Dubins trajectory is now made up of a five-segment trajectory consisting of initial, intermediate and final arc manoeuvres, together with two straight-line segments. The initial and final arc manoeuvres are modified to make the line segments tangent to the obstacle circle. There are also two possible solutions: a positive and negative rotation arc around the obstacle. Hence two avoidance trajectories are required to be calculated for every obstacle. Two possibilities occur for this case: the obstacle intersects the straight-line segment of the

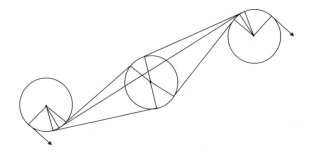

Figure 4.2   Single-obstacle Dubins

Dubins trajectory, as shown in Figure 4.2, or the obstacle intersects the initial or final arcs. We shall consider each case and the avoidance solution.

4.2.1 Line Intersection Detection

For the case where the obstacle intersects the straight-line segment, the geometry is shown in Figure 4.3. The initial calculation required is to determine whether the obstacle intersects the line segment. To do this, the distance from the centre of the obstacle to the start vertex of the line segment (d_o) is calculated, as is the projection normal to (d_n). For the line segment to intersect the obstacle, we have

$$\frac{d_n}{r_o} \leq 1, \qquad d_n \kappa_o \leq 1. \tag{4.2}$$

where r_o is the radius and κ_o is the curvature of the obstacle. If detected, the intersection of the obstacle with the line (d_i) can then be determined using the cosine rule as

$$\frac{1}{\kappa_o^2} = d_o^2 + d_i^2 - 2 d_o d_i \cos(\theta_o) \tag{4.3}$$

or

$$d_i^2 - 2 d_o \cos(\theta_o) d_i + \left(d_o^2 - \frac{1}{\kappa_o^2} \right) = 0. \tag{4.4}$$

This quadratic in d_i admits two solutions, given by

$$d_i = d_o \cos(\theta_o) \pm \sqrt{d_o^2 [\cos^2(\theta_o) - 1] + \frac{1}{\kappa_o^2}}. \tag{4.5}$$

The solutions are thus

$$d_{i_p} = d_o \cos(\theta_o) + \sqrt{d_o^2[\cos^2(\theta_o) - 1] + \frac{1}{\kappa_o^2}},$$

$$d_{i_m} = d_o \cos(\theta_o) - \sqrt{d_o^2[\cos^2(\theta_o) - 1] + \frac{1}{\kappa_o^2}}. \qquad (4.6)$$

These solutions give intersection points p_{i_p} and p_{i_m} in Figure 4.3 respectively. The first intersection point p_{i_m} is required for this application, as it represents the closest intersection point to the start vertex, hence the d_{i_m} solution is chosen. For the point to be real-valued, we also have

$$d_o^2[\cos^2(\theta_o) - 1] + \frac{1}{\kappa_o^2} \geq 0. \qquad (4.7)$$

This can be rearranged into the form

$$\frac{1}{\kappa_o^2} + d_o^2[\cos^2(\theta_o) - 1] \geq 0,$$

$$\frac{1}{\kappa_o^2} - d_o^2 \sin^2(\theta_o) \geq 0, \qquad (4.8)$$

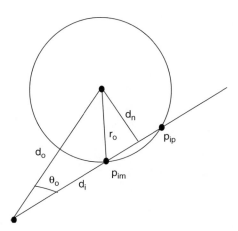

Figure 4.3   Single-obstacle Dubins line intersection

or

$$\frac{1}{\kappa_o} \geq d_o \sin(\theta_o) \geq d_n,$$

$$d_n \kappa_o \leq 0. \tag{4.9}$$

This condition is the same as that in equation (4.2) and hence detecting an intersection guarantees a real-valued solution. The intersection point will be useful in discriminating between obstacles when multiple intersections are considered.

On detection of an obstacle, two new Dubins trajectories need to be computed. This involves defining a tangent line with the preceding arc segment that is attached to the line start vertex and the obstacle circle. This is shown in Figure 4.4. Given the length of the distance between obstacle circle centre c_o and the manoeuvre arc centre c_s, simple use of Pythagoras's theorem gives the length of the tangent line d_t as

$$d_t^2 = d_c^2 - (r_o - r_s)^2, \tag{4.10}$$

$$d_t^2 = d_c^2 - (r_o + r_s)^2, \tag{4.11}$$

where equation (4.10) applies if the rotation direction of the preceding segment and the rotation direction around the obstacle are in the same direction, and equation (4.11) applies if the directions oppose. The same direction rotation case in Figure 4.4 is anticlockwise for the preceding segment and

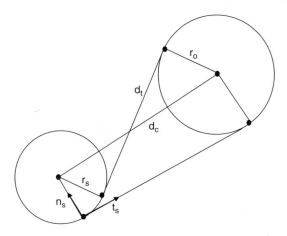

Figure 4.4 Single-obstacle Dubins line solution

anticlockwise for the obstacle traverse, and the opposing rotation case is anti-clockwise for the preceding segment and clockwise for the obstacle traverse. Defining an axis set that comprises a tangent, normal and binormal vector as shown in the figure, a positive rotation is given by an anticlockwise rotation.

For a real-valued solution to both equations, we require that

$$d_c \geq |r_o - r_s|, \tag{4.12}$$

$$d_c \geq |r_o + r_s|. \tag{4.13}$$

The equality conditions, where $d_t = 0$, are illustrated in Figures 4.5 and 4.6. Figure 4.7 shows a condition where an opposed solution is invalid but the same rotation solution is possible because the condition in equation (4.12) is satisfied. Hence, the conditions for detecting an intersection are first to test for the obstacle circle intersecting the line segment given by the condition

$$d_n \kappa_o \leq 1. \tag{4.14}$$

If an intersection is detected, the test for a valid solution for the same and opposed rotation solutions takes the form

$$d_c \geq |r_o - r_s|, \tag{4.15}$$

$$d_c \geq |r_o + r_s|. \tag{4.16}$$

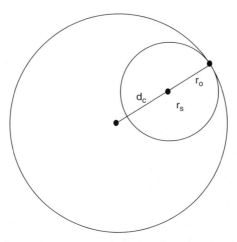

Figure 4.5 Single-obstacle Dubins line solution limit condition for the same rotation direction

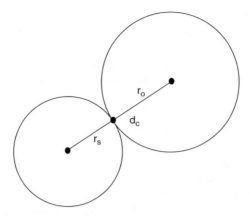

Figure 4.6 Single-obstacle Dubins line solution limit condition for opposed rotation direction

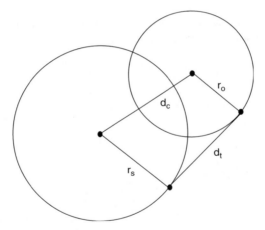

Figure 4.7 Single-obstacle Dubins intersecting line solution with the same rotation direction

4.2.2 Line Segment Intersection

Given that an intersection has been detected, two conditions exist. The first is that the valid solution conditions are met and a solution can be calculated. The second is that the valid solution conditions are violated. This second condition requires solution scaling, in that the curvature of the previous segment arc needs to be scaled to produce a valid solution. We shall consider both cases.

In order to compute the solution for an intersecting obstacle, the tangent line needs to be calculated. This entails computing its length d_t and also its direction, represented by the unit vector t_t. Figure 4.7 shows the solution with vectors defined for the radii of the arc segment r_s and the obstacle r_o, together with unit vectors defining the directions of the centre vector t_c and the tangent vector t_t. A vector summation yields

$$d_t t_t = -r_s + d_c t_c + r_o$$

$$= d_c t_c + (r_o - r_s). \tag{4.17}$$

The radii vectors are normal to the tangent vector, and hence define a tangent normal vector n_t, to give

$$r_s = \pm \frac{n_t}{\kappa_s},$$

$$r_o = \pm \frac{n_t}{\kappa_o}. \tag{4.18}$$

The rotation direction gives the required sign for the above equations. Define the rotation direction by using the variable ε, such that

$$\varepsilon = 1, \qquad \text{anticlockwise,}$$

$$\varepsilon = -1, \qquad \text{clockwise.} \tag{4.19}$$

Hence, we have

$$r_s = -\varepsilon_s \frac{n_t}{\kappa_s},$$

$$r_o = -\varepsilon_o \frac{n_t}{\kappa_o}. \tag{4.20}$$

The solution is thus of the form

$$d_t t_t = d_c t_c + [\varepsilon_s/\kappa_s - \varepsilon_o/\kappa_o]n_t \tag{4.21}$$

or

$$(\begin{matrix} t_t & n_t \end{matrix}) \begin{pmatrix} d_t \\ [\varepsilon_s/\kappa_s - \varepsilon_0/\kappa_0] \end{pmatrix} = d_c t_c. \tag{4.22}$$

As the unknown in this equation is the tangent vector t_t, the equation can be rewritten in the form

$$\begin{pmatrix} d_t & -[\varepsilon_s/\kappa_s - \varepsilon_0/\kappa_0] \\ [\varepsilon_s/\kappa_s - \varepsilon_0/\kappa_0] & d_t \end{pmatrix} t_t = d_c t_c, \tag{4.23}$$

which, in turn, can be written as

$$\begin{aligned} t_t &= d_c \begin{pmatrix} d_t & -[\varepsilon_s/\kappa_s - \varepsilon_0/\kappa_0] \\ [\varepsilon_s/\kappa_s - \varepsilon_0/\kappa_0] & d_t \end{pmatrix}^{-1} t_c \\ &= \frac{1}{d_c} \begin{pmatrix} d_t & [\varepsilon_s/\kappa_s - \varepsilon_0/\kappa_0] \\ -[\varepsilon_s/\kappa_s - \varepsilon_0/\kappa_0] & d_t \end{pmatrix} t_c, \end{aligned} \tag{4.24}$$

with the condition that the matrix is non-singular. This is true for $d_c \neq 0$, as the determinant D of the matrix is given by

$$\begin{aligned} D &= d_t^2 + \left[\frac{\varepsilon_s}{\kappa_s} - \frac{\varepsilon_0}{\kappa_0} \right]^2 \\ &= d_c^2. \end{aligned} \tag{4.25}$$

The magnitude of the tangent vector can also be rewritten in the form

$$d_t^2 = d_c^2 - \left[\frac{\varepsilon_s}{\kappa_s} - \frac{\varepsilon_0}{\kappa_0} \right]^2. \tag{4.26}$$

Equation (4.26) is used to compute the length of the tangent line segment d_t and equation (4.24) is then used to compute the tangent unit vector t_t.

The test for a valid solution can now be written in the form

$$d_c \geq \left[\frac{\varepsilon_s}{\kappa_s} - \frac{\varepsilon_0}{\kappa_0} \right]. \tag{4.27}$$

If this fails, then the arc segment curvature can be scaled to allow a valid solution. The fail cases are illustrated in Figures 4.8 and 4.9. The fail solution in fact should not occur, as any line segment tangent to the arc segment

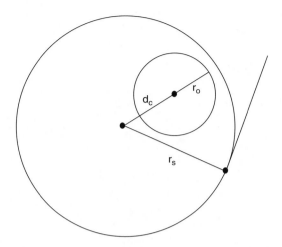

Figure 4.8   Fail condition for same rotation direction

will not intersect. Hence this case can be dismissed. The opposed rotation case in Figure 4.9 can occur and will require the arc segment curvature to be adjusted to produce the solution illustrated in Figure 4.6. If the curvature of the arc segment is adjusted, the arc centre will move, as it is defined relative to the fixed vertex at the end of the previous straight-line segment or the start vertex. This is shown in Figure 4.10. If a scaling factor s_s is applied to the curvature κ_s, then we can write, using the cosine rule,

$$\left(\frac{1}{s_s\kappa_s} + \frac{1}{\kappa_o}\right)^2 = d_p^2 + \frac{1}{s_s^2\kappa_s^2} - 2\frac{d_p}{s_s\kappa_s}\cos(\theta), \tag{4.28}$$

where d_p is the length of the vector from the fixed vertex to the obstacle centre and θ is the angle between the d_p vector and the fixed vertex normal vector. Solving for the adjusted curvature yields

$$\frac{1}{s_s\kappa_s} = \frac{d_p^2 - (1/\kappa_o^2)}{2[(1/\kappa_o) + d_p\cos(\theta)]} \tag{4.29}$$

and hence

$$s_s = \frac{2[(1/\kappa_o) + d_p\cos(\theta)]}{\kappa_s[d_p^2 - (1/\kappa_o^2)]}. \tag{4.30}$$

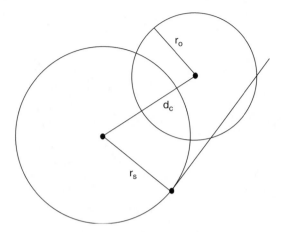

Figure 4.9   Fail condition for opposed rotation direction

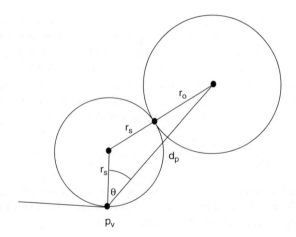

Figure 4.10   Solution for scaling start arc segment

4.2.3 Arc Intersection

The intersection of an obstacle with an arc segment is shown in Figure 4.11. The necessary condition for intersection is that the distance from the arc centre to the obstacle centre must be such that the obstacle and arc segment

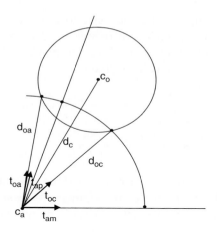

Figure 4.11   Arc intersection

intersect. Hence, for intersection,

$$d_{cm} = d_c - \frac{1}{\kappa_o} - \frac{1}{\kappa_s} < 0,$$

$$d_{cp} = d_c + \frac{1}{\kappa_o} - \frac{1}{\kappa_s} > 0, \tag{4.31}$$

where κ_s is the curvature of the arc segment and κ_o is the curvature of the obstacle. If this condition is met, the sufficient condition for intersection within the arc segment boundary needs to be evaluated. The lines from the arc segment centre c_a to the intersection point with the obstacle are defined by the anticlockwise unit vector t_{oa} and the clockwise unit vector t_{oc} with magnitude $1/\kappa_s$. The angle between the obstacle intersect vectors t_{oa} and t_{oc} and the arc radii boundary vectors t_{am} and t_{ap} are calculated from the vector cross-products and scalar products as

$$\sin(\theta_{ij}) = t_{ai} \times t_{oj},$$

$$\cos(\theta_{ij}) = t_{ai} \cdot t_{oj}, \tag{4.32}$$

$$\theta_{ij} = \frac{\sin(\theta_{ij})}{\cos(\theta_{ij})},$$

where

$$i = \begin{bmatrix} p & m \end{bmatrix}$$

$$j = \begin{bmatrix} a & c \end{bmatrix}. \tag{4.33}$$

The angles can be computed by reference to the vector connecting the arc segment centre to the obstacle centre defined by the unit vector t_c and distance d_c. The obstacle intersect vectors can then be calculated as shown in Figure 4.12. Using the cosine rule, we have

$$\frac{1}{\kappa_o^2} = d_c^2 + \frac{1}{\kappa_s^2} - 2\frac{d_c}{\kappa_s}\cos(\theta_o),$$

$$\cos(\theta_o) = \frac{d_c^2 + 1/\kappa_s^2 - 1/\kappa_o^2}{2d_c/\kappa_s}, \tag{4.34}$$

where d_c is the distance from the arc centre to the obstacle centre.

For the obstacle to intersect the arc segment, at least one of the obstacle intersect vectors must lie between the arc radii tangent vectors. This is tested by the condition

$$\theta_{pj}\theta_{mj} \le 0. \tag{4.35}$$

To illustrate this, Figure 4.13 shows a geometry that does not meet the necessary condition, whereas the geometry in Figure 4.11 has a valid intersection point.

Once detected, the solution for an intersecting arc segment is to scale the arc. Two solutions are determined. These solutions produce an arc

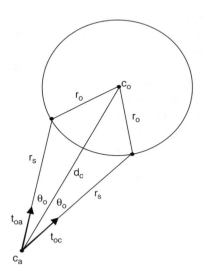

Figure 4.12   Arc intersection tangent vectors

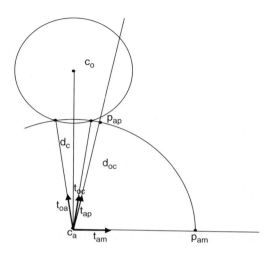

Figure 4.13 Arc intersection sufficient conditions

curvature that makes C^2 contact with the obstacle. The solutions are shown in Figures 4.14 and 4.15. To compute the arc segment radius $r_s = 1/\kappa_s$, the cosine rule can be applied. This gives

$$\left(\frac{1}{\kappa_s^2} + \frac{1}{\kappa_o^2}\right) = \frac{1}{\kappa_s^2} + d_o^2 - 2\frac{d_o}{\kappa_s}\cos(\theta_o), \qquad (4.36)$$

$$\left(\frac{1}{\kappa_s^2} - \frac{1}{\kappa_o^2}\right) = \frac{1}{\kappa_s^2} + d_o^2 - 2\frac{d_o}{\kappa_s}\cos(\theta_o), \qquad (4.37)$$

where equation (4.36) is the maximum-curvature solution shown in Figure 4.14 and equation (4.37) is the minimum-curvature solution shown in Figure 4.15. Solving for the arc segment curvature κ_s yields

$$\kappa_{s_{max}} = 2\frac{[d_o\cos(\theta_o) + 1/\kappa_o]}{d_o^2 - 1/\kappa_o^2}, \qquad (4.38)$$

$$\kappa_{s_{min}} = 2\frac{[d_o\cos(\theta_o) - 1/\kappa_o]}{d_o^2 - 1/\kappa_o^2}, \qquad (4.39)$$

where equation (4.38) is the maximum-curvature solution shown in Figure 4.14 and equation (4.39) is the minimum-curvature solution shown in Figure 4.15.

It should be noted that, while the minimum-curvature solution will produce a Dubins trajectory that avoids the obstacle, the minimum-curvature

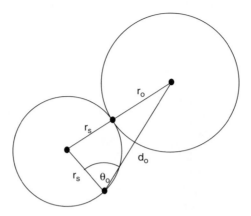

Figure 4.14   Arc scaling for maximum-curvature intersection

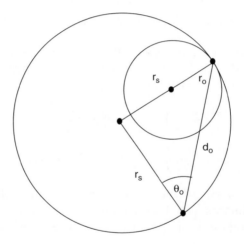

Figure 4.15   Arc scaling for minimum-curvature intersection

solution will produce a trajectory that intersects the obstacle along its line segment. The line segment solution will then give a set of two solutions with the scaled curvature that avoid the obstacle. Hence, for the intersecting arc case, three solutions will result. This is shown in Figure 4.16. For the case where multiple obstacles are present in the environment, the computation of the Dubins trajectories that avoid such obstacles is best executed by iteration. The basic Dubins trajectory that connects the start and final positions could intersect several obstacles. This is shown in Figure 4.17.

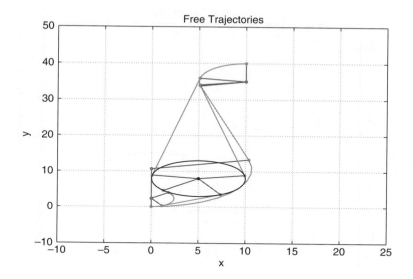

Figure 4.16   Solution set for arc intersection

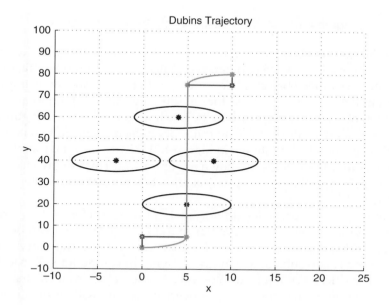

Figure 4.17   Multiple obstacle environment

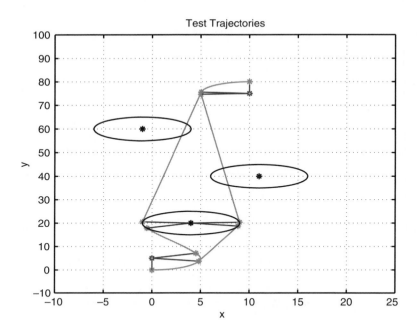

Figure 4.18   Multiple obstacle intersection trajectories

Another possibility is shown in Figure 4.18, where the trajectories generated by the basic obstacle avoidance calculations will produce trajectories that intersect other obstacles. Consider the left-hand clockwise trajectory shown in Figure 4.19. This trajectory is subject to modification to avoid the upper left obstacle, which results in the two trajectories shown in Figure 4.20. This is repeated for the right-hand trajectory, which results in two more trajectories, one of which is shown in Figure 4.21. Hence the total trajectory set comprises four trajectories, as shown in Figure 4.22. This iterative technique can be applied to the resulting trajectory set if more obstacle intersections are detected.

Note that, if any obstacles do not intersect the trajectories, they are not included in the generation of avoidance trajectories. Hence, in an obstacle-rich environment, only the obstacles that intersect the trajectory set are considered. This results in a significant saving in computational load compared with such techniques as Voronoi. It must also be noted that there is an assumption that no obstacles overlap, as this will result in one of the avoidance trajectories passing through the obstacle overlap region. The technique is efficient in that the set of trajectories shown in Figure 4.23 with

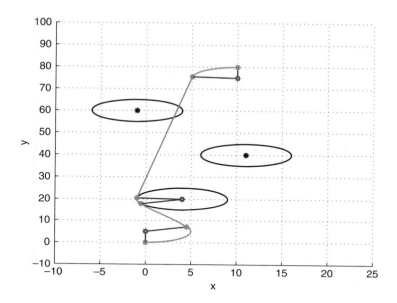

Figure 4.19   Clockwise Dubins trajectory

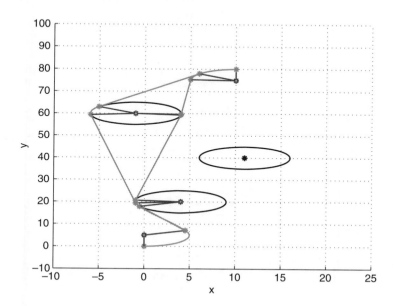

Figure 4.20   Multiple obstacle multiple trajectories for clockwise trajectory

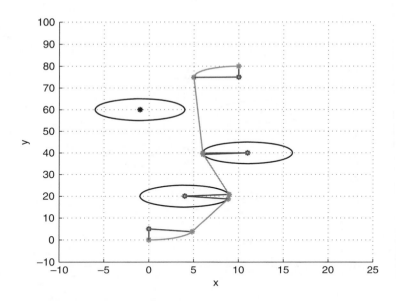

Figure 4.21   Multiple obstacle concave trajectory for anticlockwise trajectory

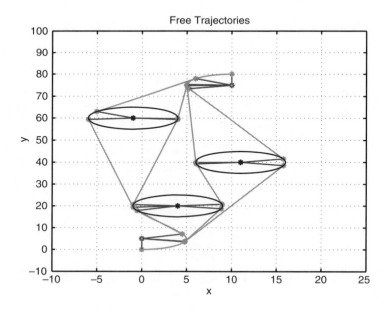

Figure 4.22   Multiple obstacle complete trajectory set

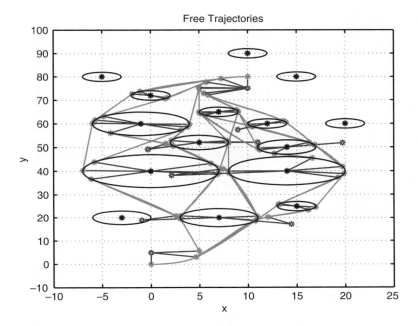

Figure 4.23 Complete trajectory set for 15 obstacles

15 obstacles resulted in a set of trajectories that avoided 10 of them. This took approximately 1 s in computing time in MATLAB code.

4.3 Obstacle Avoidance of Unmapped Static Obstacles

If a UAV detects an unmapped obstacle by means of an on-board sensor, the UAV has to re-plan the path either by varying the curvature between the two waypoints under consideration or by re-planning the path using an intermediate waypoint. The obstacles are tested by determining if the obstacle safety circle intersects the UAV safety circle. If the intersection is not empty, then re-planning is required. This can be done either by increasing the curvature of the path or by creating an intermediate waypoint and producing a new path that includes this new waypoint.

Assuming the obstacle circle O_{obs} and the UAV safety circle O_{safe}, the condition for collision avoidance is

$$O_{obs} \cap O_{safe} = \varnothing. \tag{4.40}$$

4.3.1 Safety Circle Algorithm

Assuming that a group of UAVs are cooperating in the manner described in Chapter 6, then the path planning algorithm described in that chapter needs to be modified. A simple extension of the algorithm to modify the paths of the UAVs by changing the curvature of the arcs of the path is given as follows:

   (i) Produce flyable paths for each UAV.
  (ii) Change the course of the path to meet the safety constraints by increasing the curvature.
 (iii) Calculate the length of the paths.
  (iv) Find the reference path.
   (v) Increase the length of the shorter paths to the length of the reference path. This results in paths of equal length.
  (vi) Check again for the paths meeting the safety constraints.
 (vii) If not, adjust the position of the new waypoint and increase the curvature to meet the safety conditions to produce paths of equal length.

Although the above algorithm can be used to ensure collision avoidance, handling unmapped obstacles is best achieved by generating intermediate waypoints and/or poses that will avoid the obstacle, as this algorithm is faster. This is because it does not depend on an iterative process of increasing the curvature until an avoidance path is produced. The advantages of this method is that it is efficient and implementation is simple.

4.3.2 Intermediate Waypoint Algorithm

When an obstacle intersection is detected, the path planner generates an intermediate waypoint so that the obstacle is avoided. Consider a flyable trajectory $r(t)$ generated for a given set of poses and/or waypoints. The obstacle avoidance algorithm calls the path planner to re-plan the nominal path by selecting a new waypoint or pose. The schematic of the concept is given in Figure 4.24. In the figure, the central hatched circle is the obstacle. Points M and N on the safety circle are the intermediate poses, and the points of intersection of the obstacle safety circle and the nominal path (dotted in the figure) are given by X_1 (entry) and X_2 (exit). The intermediate waypoints are generated by first drawing a line between the entry point X_1 and the exit point X_2. If a line orthogonal to this a constructed, it will intersect the obstacle safety circle at two point N and M. These are then designated as the potential intermediate waypoints. The intermediate waypoint is selected based on the

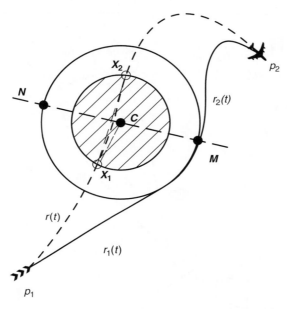

Figure 4.24 Threat handling by intermediate pose. Reprinted with permission of Elsevier

location of the centre of the obstacle C, which is either to the left or the right of this line. If the centre C is left to the line X_1-X_2, the intermediate pose M is selected on the right-hand side of the obstacle region, and vice versa. In the figure, the initial dotted path is $r(t)$ and the modified solid line paths, generated with the intermediate waypoint M, are $r_1(t)$ and $r_2(t)$. Hence the multiple UAV path planning algorithm developed in Chapter 6 is modified to give the following:

   (i) Produce flyable paths for each UAV.
  (ii) Change the course of the path to meet the safety constraints by creating intermediate waypoints.
 (iii) Calculate the length of the paths.
 (iv) Find the reference path.
  (v) Increase the length of the shorter paths to the length of the reference path. This results in paths of equal length.
 (vi) Check again for the paths meeting the safety constraints.
(vii) If not, adjust the position of the new waypoint and increase the curvature to meet the safety conditions to produce paths of equal length.

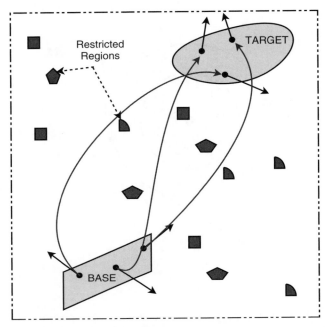

Figure 4.25 UAVs in a static cluttered environment. Reprinted with permission of ASME

4.4 Algorithmic Implementation

A schematic for a cluttered environment is shown in Figure 4.25. The obstacle regions are modelled as polygons but will have safety circles associated with them. The regions are assumed stationary and their positions are known. The path planning of multiple UAVs through a sequence of poses can be written as

$$P_{s,i,j-1}(x_{s,i,j-1}, y_{s,i,j-1}, \theta_{s,i,j-1}) \xrightarrow{r_{i,j-1}(q)} P_{f,i,j}(x_{f,i,j}, y_{f,i,j}, \theta_{f,i,j}), \tag{4.41}$$

$$i = 1, \ldots, n_{\text{UAV}}, \quad j = 2, \ldots, n_{\text{p}}, \quad |\kappa_i(q)| < \kappa_{\max}, \quad \bigcup_{\text{safe}}, \bigcup_{\text{length}},$$

where n_{UAV} is the number of UAVs, n_{p} is the number of poses (including the start and finish poses), the suffix i is for the ith UAV and the suffix j is for the jth pose. Two UAVs are considered for implementation of the algorithm. The UAVs are assumed to be homogeneous in their physical capabilities, and hence they will both fly at the same speed and have the

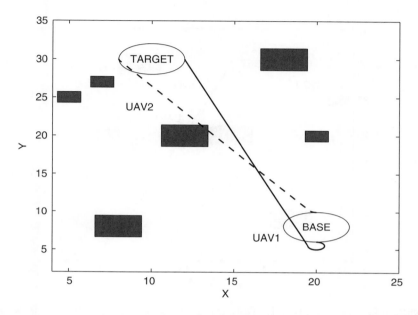

Figure 4.26   Dubins flyable paths of two UAVs in a cluttered environment

same curvature constraints. In previous chapters, Dubins, clothoid and PH paths are generated, so we shall consider solutions based on each of these.

4.4.1 Dubins Path Modification

Figure 4.26 shows the paths of two UAVs in a cluttered environment. The flights path of UAV2 intersects the obstacle and with the flight path of UAV1. The curvatures of the circular arcs of the Dubins path are varied till the flight path avoids the threat region. Figure 4.27 shows the new safe and flyable path after increasing the curvature of the UAV2 path. Figure 4.28 shows the solution of the same problem by using an intermediate waypoint. Once the obstacle intersection has been detected, an intermediate waypoint is generated using the principle explained in section 4.3.2.

4.4.2 Clothoid Path Modification

Figure 4.29 shows nominal paths as a result of using clothoid paths. The poses and threats are generated randomly. The distances between the poses are at least twice as great as the minimum radius of curvature (maximum curvature) of the UAV. This is to ensure that a flyable path exists between the

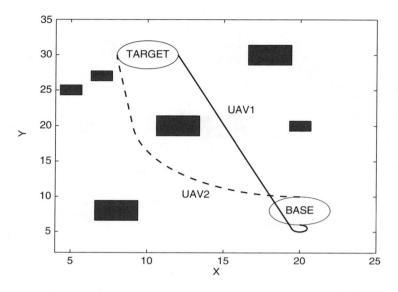

Figure 4.27   Re-planning the Dubins path of UAV2 by curvature adjustment

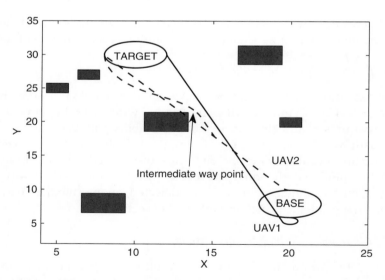

Figure 4.28   Re-planning the Dubins path of UAV2 using an intermediate waypoint

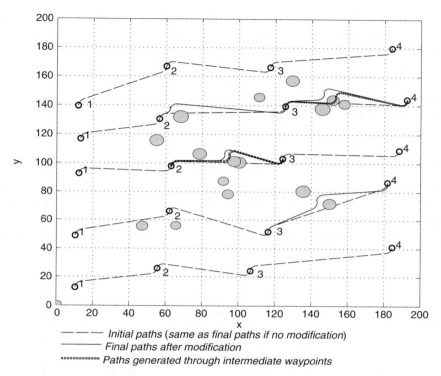

Initial paths (same as final paths if no modification)
Final paths after modification
Paths generated through intermediate waypoints

Figure 4.29 Five UAVs each with four waypoints in cluttered space. Reprinted with permission of Elsevier

poses. The filled ellipses are the obstacles. Flyable paths are then produced between the poses to give the nominal paths (dashed in the figure). The obstacle intersections are handled by generating intermediate waypoints and/or poses as explained in section 4.3.2. The intersection of paths with the no-fly zones are determined iteratively for the clothoid arc segments, while the intersection of lines with ellipses can be detected by simple geometry. An intermediate pose or waypoint is produced for each intersection within the obstacle safety zone. Following this, the paths are re-planned to pass through the new waypoints.

From the figure, it is also seen that, when a path intersects with a single obstacle, re-planning of a new path is simple, as it only has to avoid a single obstacle. However, re-planning is difficult when obstacles appear in clusters, as in the case of the second and third UAVs. In the case of the second UAV, and in between the third and fourth waypoints, there are three obstacles in a cluster. The first re-planning of the initial path intersects the other obstacle and hence requires further re-planning to avoid all of the obstacles. A similar situation arises for the third UAV, where a cluster of two obstacles appear together. Finally, the curvature of the first four paths, ordered from the top, are increased to that of the reference path, which is the path length of fifth UAV. This is similar to the path planning for a cluttered environment for mapped obstacles detailed in section 4.2 using the differential geometric approach.

4.4.3 PH Path Modification

For the case of PH paths, first the initial paths have to be optimized for curvature continuity, as the initial paths are only tangent-continuous. The flight paths shown in Figure 4.30 are the initial nominal paths, which are tangent-continuous with the start pose P_s and the finish pose P_f. As the paths

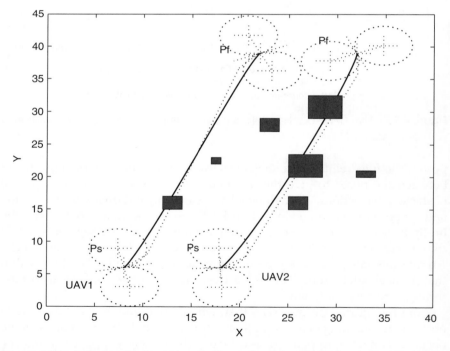

Figure 4.30 Initial paths (only tangent continuity) – PH 2D in cluttered space. Reprinted with permission of ASME

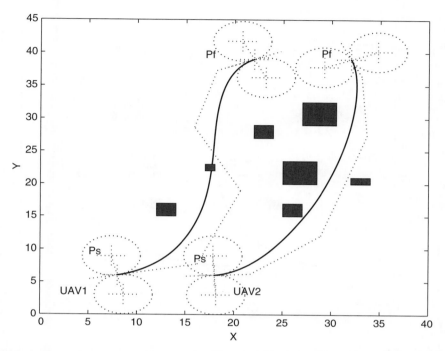

Figure 4.31 Flyable paths – PH 2D in cluttered space. Reprinted with permission of ASME

do not meet the maximum-curvature bound, they have to be optimized first for the flyable path. The resulting path length of UAV1 is 35.96 units and that of UAV2 is 35.92 units.

Figure 4.31 shows the UAV paths optimized for their curvatures. The resulting lengths of the new paths are 40.69 and 37.13 units respectively. Now the resulting paths do not meet the safety requirements, as intersections with the obstacle safety circles and the UAV safety circles occur. The path of UAV1 intersects with the obstacles, while that of UAV2 has a safety circle intersection. Both the paths need further modification to their curvature.

Further increase in the curvature of the paths result in paths that meet the safety constraints. Figure 4.32 shows the resulting safe flight paths. The corresponding path lengths are now 42.57 and 41.12 units. As the length s_1 is longer than s_2, the flight path r_1 is designated as the reference path. The length of r_2 thus has to be increased to that of r_1 to produce paths of equal lengths. Figure 4.33 shows the final paths of equal length 42.57 units.

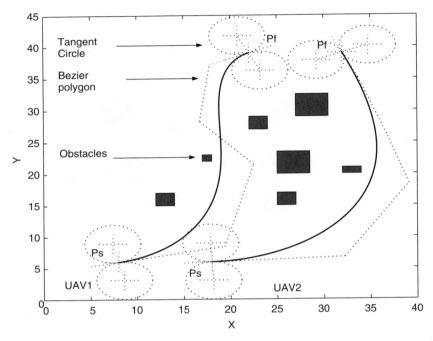

Figure 4.32 Feasible (safe and flyable) paths – PH 2D in cluttered space. Reprinted with permission of ASME

4.4.4 Obstacle Avoidance in 3D

For obstacle avoidance in 3D, a simple case of flying through an urban area containing buildings is considered. Each building is square in shape, and the size of each building is different. The buildings are overlapping in this example because the locations of the buildings are generated randomly. This is not as artificial as it first appears, because urban environments are complex and building shapes include such intersecting squares as representative shapes. The scenario is shown in Figure 4.34. The heights and areas of the buildings are generated randomly. As shown in the figure, each building is enclosed by a cylinder of diameter equal to the length of the diagonal of the square. This is to simplify the distance calculation between the buildings. If the distance between the buildings is less than a threshold value, the UAV cannot fly between them.

The positions of the obstacles are defined by means of the centres of the encompassing safe distance circles. Delaunay triangulation (Berg *et al.* 2000)

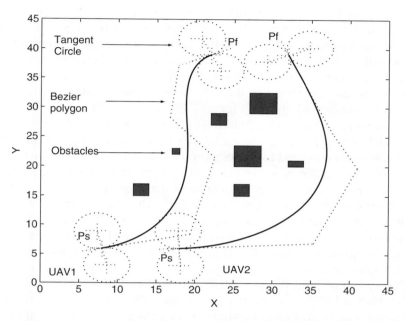

Figure 4.33 Paths of equal lengths – PH 2D in cluttered space. Reprinted with permission of ASME

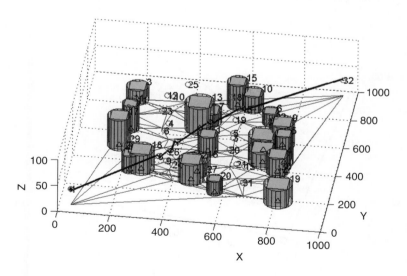

Figure 4.34 Obstacle avoidance in 3D

is applied on this database, which connects each point in the database to its natural neighbours, thus forming a connectivity graph of obstacles to their immediate neighbours. The UAV has to fly across these connecting lines between the obstacles. In order to make sure that the distances between the obstacles are large enough for the UAV, any connecting lines between the obstacles whose length is less than the safety radius of the UAV is deleted from the connectivity graph. Figure 4.35 shows the 2D projection of the triangulation of 10 obstacles (buildings). The figure shows the connectivity between their neighbourhoods – for example, the obstacle no. 9 at the bottom right corner is connected to its neighbours 1, 4 and 8, but not to any others. This helps to decide whether the UAV can fly in-between the obstacles. This is done by calculating the available distance between the obstacles, considering the size (radius) of the buildings. The small open circles are the centre points (nodes) of the lines through which the UAV can fly. The nodes are waypoints for the UAVs. Route planning is done by finding the shortest route through these nodes. The Dijkstra search algorithm (Dijkstra 1959) is used to find the route.

The waypoints can be connected using either Dubins, clothoid, or PH paths. Here a Dubins path is used to connect the nodes. The height of the waypoints can be chosen depending on the possible climb rate of the UAV. The simulations in Figures 4.34 and 4.36 use 30 obstacles. The grey lines are the triangulation of obstacle positions, and the thin black lines are the

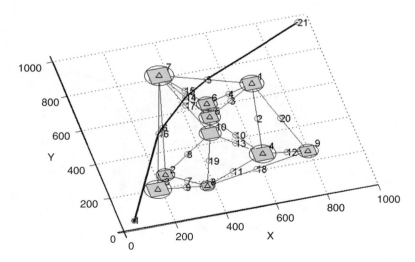

Figure 4.35   Path planning with obstacle avoidance in 3D

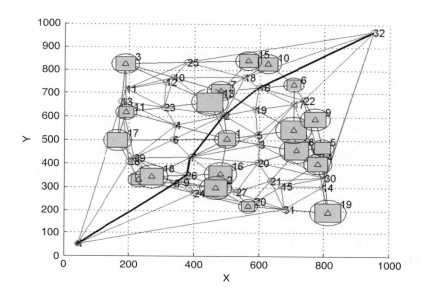

Figure 4.36   Obstacle avoidance in 3D (2D projection)

triangulation formed by the nodes including the start point (node no. 1) and goal point (node no. 32). The numbers on the buildings are the label numbers for each building.

References

Agarwal, P. K., Biedl, T., Lazard, S., Robbins, S., Suri, S. and Whitesides, S. 2002. Curvature-constrained shortest paths in a convex polygon. *SIAM Journal of Computing*, **31**(6), 1814–1851.

Assaf, D. and Sharlin-Bilitzky, A. 1994. Dynamic search for a moving target. *Journal of Applied Probability*, **31**(2), 438–457.

Beard, R., McLain, T., Goodrich, M. and Anderson, E. 2002. Coordinated target assignment and intercept for unmanned air vehicles. *IEEE Transactions on Robotics and Automation*, **18**(6), 911–922.

Benkoski, S. J., Monticino, M. G. and Weisinger, J. R. 1991. A survey of the search theory literature. *Naval Research Logistics*, **38**(4), 468–494.

Berg, M., Kreveld, M., Overmars, M. and Schwarzkopf, O. 2000. *Computation Geometry, Algorithms and Applications*. Springer.

Bicchi, A. and Pallottino, L. 2000. An optimal cooperative conflict resolution for air traffic management systems. *IEEE Transactions on Intelligent Transportation Systems*, **1**(4), 221–232.

Bortoff, S. 2000. Path-planning for unmanned air vehicles. *Proc. American Control Conf.*, pp. 364–368.

Chan, Y. K. and Foddy, M. 1985. Real time optimal flight path generation by storage of massive data bases. *Proc. IEEE NEACON 1985*, pp. 516–521. Institute of Electrical and Electronics Engineers.

Dijkstra, E. W. 1959. A note on two problems in connexion with graphs. *Numerische Mathematik*, 1, 269–271.

Dowek, G., Muñoz, C. and Geser, A. 2001. Tactical conflict detection and resolution in a 3-D airspace. Technical Report, ICASE Report No. 2001-7, NASA/CR-2001-210853.

Eagle, J. N. and Yee, J. R. 1990. An optimal branch-and-bound procedure for the constrained path, moving target search problem. *Operations Research*, 38(1), 110–114.

Eun, Y. and Bang, H. 2006. Cooperative control of multiple unmanned aerial vehicles using the potential field theory. *Journal of Aircraft*, 43(6), 1805–1814.

Fujimori, A., Ogawa, Y. and Nikiforuk, P. N. 2002. A modification of cooperative collision avoidance for multiple robots using the avoidance circle. *Proceedings of the Institution of Mechanical Engineers, Part I, Journal of Systems and Control Engineering*, 216(3), 291–299.

Hebert, J., Jacques, D., Novy, M. and Pachter, M. 2001. Cooperative control of UAVs. *AIAA Guidance, Navigation, and Control Conf. and Exhibit*, AIAA-2001-4240.

Kim, J. O. and Khosla, P. K. 1992. Real-time obstacle avoidance using harmonic potential functions. *IEEE Transactions on Robotics and Automation*, 8, 338–349.

Kitamura, Y., Tanaka, T., Kishino, F. and Yachida, M. 1995. 3-D path planning in a dynamic environment using an octree and an artificial potential field. *Proc. IEEE/RSJ Int. Conf. on Intelligent Robots and Systems*, vol. 2, pp. 474–481.

Kitamura, Y., Tanaka, T., Kishino, F. and Yachida, M. 1996. Real-time path planning in a dynamic 3-D environment. *Proc. IEEE/RSJ Int. Conf. on Intelligent Robots and Systems*, vol. 2, pp. 925–931.

Koopman, B. O. 1980. *Search and Screening: General Principles with Historical Applications*. Pergamon Press.

Lamiraux, F., Bonnafous, D. and Lefebvre, O. 2004. Reactive path deformation for nonholonomic mobile robots. *IEEE Transactions on Robotics*, 20(6), 967–977.

Li, S. M., Boskovic, J. D., Seereeram, S., Prasanth, R., Amin, J., Mehra, R. K., Beard, R. and McLain, T. W. 2002. Autonomous hierarchical control of multiple unmanned combat air vehicles (UCAVs). *American Control Conf.*, Anchorage, AK, vol. 1, pp. 274–279.

Lozano-Pérez, T. 1983. Spatial planning: a configuration space approach. *IEEE Transactions on Computing*, 32(2), 108–120.

McLain, T. 2000. Cooperative rendezvous of multiple unmanned air vehicles. *Proc. AIAA Guidance and Control Conf.*, Denver.

Reif, J. and Wang, H. 1998. The complexity of the two dimensional curvature-constrained shortest path problem. *Proc. Third Workshop on the Algorithmic Foundations of Robotics, WAFR 98*, Massachusetts.

Richards, A. and How, J. P. 2002. Aircraft trajectory planning with collision avoidance using mixed integer linear programming. *American Control Conf.*, pp. 1936–1941.

Shanmugavel, M., Tsourdos, A., Żbikowski, R. and White, B. A. 2005. Path planning of multiple UAVs in an environment of restricted regions. *Proc. ASME Int. Mechanical Engineering Congress and Exposition, IMECE2005*, 5–11 November, Orlando, FL. IMECE2005-79682.

Stone, L. D. 1975. *Theory of Optimal Search*. Academic Press.

Thomas, L. C. and Eagle, J. N. 1995. Criteria and approximate methods for path-constrained moving-target search problems. *Naval Research Logistics*, **42**, 27–38.

Vian, J. L. and More, J. R. 1989. Trajectory optimization with risk minimization for military aircraft. *AIAA Journal of Guidance, Control, and Dynamics*, **12**(3), 311–317.

Washburn, A. R. 1983. Search for a moving target: the FAB algorithm. *Operations Research*, **31**, 739–751.

Washburn, A. R. 1990. Continuous autorouters, with an application to submarines. Research Report, NPSOR-91-05, Naval Postgraduate School, Monterey, CA.

Yang, H. I. and Zhao, Y. J. 2004. Trajectory planning for autonomous aerospace vehicles amid obstacles and conflicts. *Journal of Guidance, Control and Dynamics*, **27**, 997–1008.

Zabarankin, M., Uryasev, S. and Pardalos, P. 2002. Optimal risk path algorithms. In *Cooperative Control and Optimization* (eds Murphey, R. and Pardalos, P.), pp. 271–303. Kluwer Academic.

Zeitlin, A. D. and McLaughlin, M. P. 2007. Safety of cooperative collision avoidance for unmanned aircraft. *IEEE Aerospace and Electronics Magazine*, 9–13.

5

Path-Following Guidance

The previous chapters have detailed the design of paths through complex environments to enable the UAVs to coordinate their search and rendezvous activities. In order for the UAVs to complete their missions, we require them to be able to follow the planned paths accurately. To that end, guidance algorithms are required to ensure correct following of the paths. The analysis will assume that an autopilot has been designed to enable the UAV to follow velocity and heading commends accurately. Although this is not a topic for this book, there are many references that confirm that the design of such autopilots has been demonstrated in flight.

In order to fly along or close to the planned path, the UAV may be assigned to track the trajectory or to follow the path. The first case is called trajectory tracking, where the path is parameterized by time and the UAV has to track a point as it moves along the path. The second method is called path following, in which the UAV has to stay close to the path without having to track an explicit point. In both methods, the guidance system tries to reduce the distance error between the UAV and the path. The distance error is the orthogonal projection of the UAV's position onto the path. The guidance algorithm needs to generate a smooth path that meets the kinematic constraints of the UAV and to minimise the error. The guidance algorithm must also be able to make the UAV track straight paths and curved paths.

Cooperative Path Planning of Unmanned Aerial Vehicles
Antonios Tsourdos, Brian White and Madhavan Shanmugavel
© 2011 John Wiley & Sons, Ltd

Some path-following guidance algorithms guide the UAV from one waypoint to the next, but this can result in oscillations of the UAV path, which produces a 'slalom'-type trajectory rather than a straight-line trajectory. The waypoint-following control becomes difficult when the UAV location is disturbed, by wind, for example, when it results in oscillations around the waypoints. Following curved trajectories is also challenging with this method (Park *et al.* 2004).

A better way is to track the path joining the waypoints (Caravita *et al.* 2007). This path-following method ensures that the error is minimised. The path-following method was originally reported in the field of mobile robotics, where it is used to reduce the distance between the robot's position and the path, and the angle between the velocity vector and the tangent on the path. Nonlinear control for path following ensures stabilisation and convergence in mobile robots (Micaelli and Samson 1992, 1993). A Lyapunov direct method is used in Kim and Oh (1999) to prove convergence. A gain scheduling controller with linearization of the error vector about the pre-planned path is reported in Pascoal *et al.* (2006). A comprehensive review of path following for ocean vehicles is reported in Fossen (1994). A method similar to the potential field and vector field method is used for path planning of micro-UAVs in Nelson *et al.* (2007). Here we present the carrot guidance algorithm for the UAV.

Although in practical systems the UAV has an autopilot that has its own dynamics, it will be assumed that the UAV will instantaneously attain the desired turn rate required by the guidance algorithm. This is not usually a problem if the autopilot that determines the UAV response is fast compared with the dynamic response of the guidance system. A more complex analysis is required if the dynamics of the UAV, as well as the kinematics of the guidance system, are considered together. An integrated guidance and control method is used to follow waypoints in Kaminer *et al.* (1998).

5.1 Path Following the Dubins Path

Here, we consider a UAV following a Dubins path, as shown in Figure 5.1. The figure shows the UAV axes t_b and n_b, which represent the x and y axes of the UAV in body axes, with θ as the yaw angle of the UAV. There is also a set of axes that are fixed in the Dubins curve (t_c and n_c), which have an origin on the curve such that the UAV lies along the normal axis n_c. Hence, as the UAV moves in inertial space, the Dubins curve axes will move along the curve to maintain the UAV position on the normal vector n_c.

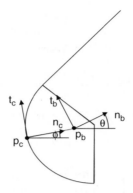

Figure 5.1   Guidance geometry

In inertial axes, we have

$$V_{\mathrm{I}} = \mathbf{R}(\theta)V_{\mathrm{b}}, \qquad (5.1)$$

where V_{b} is the UAV speed, and θ is the body axes rotation angle with respect to the inertial axes. Hence

$$\mathbf{R}(\theta) = \begin{pmatrix} \cos(\theta) & -\sin(\theta) & 0 \\ \sin(\theta) & \cos(\theta) & 0 \\ 0 & 0 & 1 \end{pmatrix},$$

$$V_{\mathrm{b}} = \begin{pmatrix} V_{\mathrm{b}} \\ 0 \\ 0 \end{pmatrix}. \qquad (5.2)$$

If a set of path axes are defined that are attached to the UAV via the normal vector n_{c}, then the UAV position in inertial axes is given by

$$p_{\mathrm{I}} = p_{\mathrm{c}} + \mathbf{R}(\phi)p_{\mathrm{u}}, \qquad (5.3)$$

where

$$\mathbf{R}(\phi) = \begin{pmatrix} \cos(\phi) & -\sin(\phi) & 0 \\ \sin(\phi) & \cos(\phi) & 0 \\ 0 & 0 & 1 \end{pmatrix},$$

$$p_{\mathrm{u}} = \begin{pmatrix} 0 \\ d \\ 0 \end{pmatrix}, \qquad (5.4)$$

where d is the line connecting the point on the Dubins curve and the UAV, $\boldsymbol{p}_u$ is the unit vector along this line and ϕ is the arc angle of the resulting axes.

Differentiating with respect to time yields

$$\frac{d\boldsymbol{p}_I}{dt} = \frac{d\boldsymbol{p}_c}{dt} + \frac{d\mathbf{R}(\phi)}{d\phi}\frac{d\phi}{dt}\boldsymbol{p}_u + \mathbf{R}(\phi)\frac{d\boldsymbol{p}_u}{dt},$$

$$\dot{\boldsymbol{p}}_I = \dot{\boldsymbol{p}}_c + \hat{\mathbf{R}}(\phi)\dot{\phi}\boldsymbol{p}_u + \mathbf{R}(\phi)\dot{\boldsymbol{p}}_u, \tag{5.5}$$

where

$$\hat{\mathbf{R}}(\phi) = \begin{pmatrix} -\sin(\phi) & -\cos(\phi) & 0 \\ \cos(\phi) & -\sin(\phi) & 0 \\ 0 & 0 & 0 \end{pmatrix}. \tag{5.6}$$

Now

$$\dot{\boldsymbol{p}}_I = V_I$$

$$= \mathbf{R}(\phi)\begin{pmatrix} V_b \\ 0 \\ 0 \end{pmatrix} \tag{5.7}$$

and

$$\dot{\boldsymbol{p}}_c = \mathbf{R}(\phi)\begin{pmatrix} V_c \\ 0 \\ 0 \end{pmatrix}. \tag{5.8}$$

We also have

$$\dot{\boldsymbol{p}}_u = \begin{pmatrix} 0 \\ \dot{d} \\ 0 \end{pmatrix}, \tag{5.9}$$

where V_c is the speed of movement of the point $\boldsymbol{p}_c$ along the path. Equation (5.5) can thus be written as

$$\mathbf{R}(\theta)\begin{pmatrix} V_b \\ 0 \\ 0 \end{pmatrix} = \mathbf{R}(\phi)\begin{pmatrix} V_c \\ \dot{d} \\ 0 \end{pmatrix} + \hat{\mathbf{R}}(\phi)\dot{\phi}\begin{pmatrix} 0 \\ d \\ 0 \end{pmatrix}. \tag{5.10}$$

Written in terms of the path axes, we have

$$\mathbf{R}'(\phi)\mathbf{R}(\theta)\begin{pmatrix} V_b \\ 0 \\ 0 \end{pmatrix} = \begin{pmatrix} V_c \\ \dot{d} \\ 0 \end{pmatrix} + \mathbf{R}'(\phi)\hat{\mathbf{R}}(\phi)\dot{\phi}\begin{pmatrix} 0 \\ d \\ 0 \end{pmatrix} \tag{5.11}$$

or

$$\mathbf{R}_V\begin{pmatrix} V_b \\ 0 \\ 0 \end{pmatrix} = \begin{pmatrix} V_c \\ \dot{d} \\ 0 \end{pmatrix} + \mathbf{R}_\phi\dot{\phi}\begin{pmatrix} 0 \\ d \\ 0 \end{pmatrix}, \tag{5.12}$$

where

$$\begin{aligned}
\mathbf{R}_V &= \begin{pmatrix} \cos(\phi) & \sin(\phi) & 0 \\ -\sin(\phi) & \cos(\phi) & 0 \\ 0 & 0 & 1 \end{pmatrix}\begin{pmatrix} \cos(\theta) & -\sin(\theta) & 0 \\ \sin(\theta) & \cos(\theta) & 0 \\ 0 & 0 & 1 \end{pmatrix} \\
&= \begin{pmatrix} \cos(\phi)\cos(\theta) + \sin(\phi)\sin(\theta) & -\cos(\phi)\sin(\theta) + \sin(\phi)\cos(\theta) & 0 \\ -\sin(\phi)\cos(\theta) + \cos(\phi)\sin(\theta) & \sin(\phi)\sin(\theta) + \cos(\phi)\cos(\theta) & 0 \\ 0 & 0 & 1 \end{pmatrix} \\
&= \begin{pmatrix} \cos(\theta - \phi) & -\sin(\theta - \phi) & 0 \\ \sin(\theta - \phi) & \cos(\theta - \phi) & 0 \\ 0 & 0 & 1 \end{pmatrix}
\end{aligned} \tag{5.13}$$

and

$$\begin{aligned}
\mathbf{R}_\phi &= \begin{pmatrix} \cos(\phi) & \sin(\phi) & 0 \\ -\sin(\phi) & \cos(\phi) & 0 \\ 0 & 0 & 1 \end{pmatrix}\begin{pmatrix} -\sin(\phi) & -\cos(\phi) & 0 \\ \cos(\phi) & -\sin(\phi) & 0 \\ 0 & 0 & 0 \end{pmatrix} \\
&= \begin{pmatrix} 0 & -1 & 0 \\ 1 & 0 & 0 \\ 0 & 0 & 0 \end{pmatrix}.
\end{aligned} \tag{5.14}$$

Hence we obtain

$$\begin{pmatrix} \cos(\theta - \phi) & -\sin(\theta - \phi) & 0 \\ \sin(\theta - \phi) & \cos(\theta - \phi) & 0 \\ 0 & 0 & 1 \end{pmatrix}\begin{pmatrix} V_b \\ 0 \\ 0 \end{pmatrix} = \begin{pmatrix} V_c \\ \dot{d} \\ 0 \end{pmatrix} + \begin{pmatrix} 0 & -1 & 0 \\ 1 & 0 & 0 \\ 0 & 0 & 0 \end{pmatrix}\dot{\phi}\begin{pmatrix} 0 \\ d \\ 0 \end{pmatrix}. \tag{5.15}$$

Expanding the equations gives

$$V_b \cos(\theta - \phi) = V_c - \dot{\phi}d,$$
$$V_b \sin(\theta - \phi) = \dot{d} \tag{5.16}$$

or

$$V_c = V_b \cos(\theta - \phi) + \dot{\phi}d, \tag{5.17}$$
$$\dot{d} = V_b \sin(\theta - \phi). \tag{5.18}$$

Equation (5.17) gives the speed of the path axes origin along the path. Equation (5.18) gives the dynamics of the distance d from the path.

Now, the rate of change of ϕ, the curvature of the path κ and the speed V_c of the point on the path p_c are related by

$$\dot{\phi} = \kappa V_c. \tag{5.19}$$

Hence

$$V_c = V_b \cos(\theta - \phi) + \kappa V_c d$$
$$(1 - \kappa d)V_c = V_b \cos(\theta - \phi)$$
$$V_c = \frac{V_b \cos(\theta - \phi)}{(1 - \kappa d)}. \tag{5.20}$$

This becomes indeterminate when the distance from the path coincides with the turn centre of the arc of the path. This occurs when

$$(1 - \kappa d) = 0$$
$$d = \frac{1}{\kappa}. \tag{5.21}$$

This is a condition where V_c becomes indeterminate. For this condition, the UAV passes through the centre of rotation of the arc of the Dubins path.

5.2 Linear Guidance Algorithm

Consider a straight-line path, where, without loss of generality, the line is at an angle given by $\phi = 0$. We also have $\dot{\phi} = 0$. Hence

$$V_c = V_b \cos(\theta),$$
$$\dot{d} = V_b \sin(\theta), \tag{5.22}$$

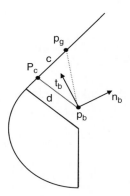

Figure 5.2 Carrot guidance

where V_b is the velocity of the UAV. The second equation requires an active control algorithm to drive d to zero. The classic form of guidance is to assume a carrot at a distance of c further along the path. This is shown in Figure 5.2.

For a zero-curvature path, we have

$$\tan \psi = \frac{d}{c}, \tag{5.23}$$

where ψ is the angle between the path and a line drawn between the carrot point and the UAV body axes origin. In order to drive the UAV onto the path, the velocity vector of the UAV needs to be controlled onto the vector of the carrot point from the body. Hence consider a feedback algorithm such that

$$\dot{\theta} = -K(\psi + \theta). \tag{5.24}$$

Bringing the equations together yields

$$\dot{d} = V \sin(\theta),$$
$$\dot{\theta} = -K \left[\tan^{-1} \left(\frac{d}{c} \right) + \theta \right]. \tag{5.25}$$

For small signals, we have

$$\dot{d} = V\theta,$$
$$\dot{\theta} = -K \left[\left(\frac{d}{c} \right) + \theta \right] \tag{5.26}$$

or, in state space form,

$$\begin{pmatrix} \dot{d} \\ \dot{\theta} \end{pmatrix} = \begin{pmatrix} 0 & V \\ -K/c & -K \end{pmatrix} \begin{pmatrix} d \\ \theta \end{pmatrix}. \tag{5.27}$$

The characteristic equation thus takes the form

$$|sI - A| = \begin{vmatrix} s & V \\ -K/c & s+K \end{vmatrix}$$

$$= s^2 + Ks + \frac{KV}{c}$$

$$= 0. \tag{5.28}$$

This is stable, with the natural frequency and damping determined by the speed V of the UAV, the carrot distance c and the loop gain K. Figure 5.3 shows the resulting trajectory of the guidance algorithm following a Dubins trajectory.

The figure shows that the velocity vector points at the carrot point in all three figures. As the line joining the carrot point to the UAV position is a straight line, the trajectory will follow the Dubins path exactly, but will exhibit an offset or error when following the curved sections of the Dubins path. In order to follow the curved sections of the trajectory, a more complex guidance algorithm is required.

5.3 Nonlinear Dynamic Inversion Guidance

When following a Dubins trajectory, the algorithm must produce a guidance trajectory that matches the curvature of the Dubins path. A suitable approach is to use nonlinear dynamic inversion as a guidance approach. This linearizes the control equations to give a defined response onto the trajectory and results in a matched curvature response. Hence, consider a desired response of the form

$$\ddot{d} + 2\zeta\omega_n\dot{d} + \omega_n^2 = 0. \tag{5.29}$$

This equation requires equations in d, $\dot{d}$ and $\ddot{d}$. To determine the second derivative of d, differentiate equation (5.5) or equivalently equation (5.10).

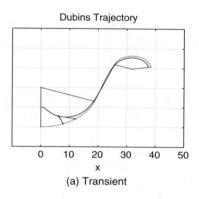

(a) Transient

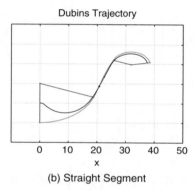

(b) Straight Segment

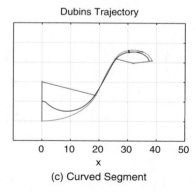

(c) Curved Segment

Figure 5.3 Linear guidance trajectory

This yields

$$\hat{\mathbf{R}}(\theta)\dot{\theta}\begin{pmatrix} V_b \\ 0 \\ 0 \end{pmatrix} = \hat{\mathbf{R}}(\phi)\dot{\phi}\begin{pmatrix} V_c \\ \dot{d} \\ 0 \end{pmatrix} + \mathbf{R}(\phi)\begin{pmatrix} \dot{V_c} \\ \ddot{d} \\ 0 \end{pmatrix} + \hat{\hat{\mathbf{R}}}(\phi)\dot{\phi}^2\begin{pmatrix} 0 \\ d \\ 0 \end{pmatrix}$$

$$+ \hat{\mathbf{R}}(\phi)\ddot{\phi}\begin{pmatrix} 0 \\ d \\ 0 \end{pmatrix} + \hat{\mathbf{R}}(\phi)\dot{\phi}\begin{pmatrix} 0 \\ \dot{d} \\ 0 \end{pmatrix}, \tag{5.30}$$

where

$$\hat{\mathbf{R}}(\theta) = \begin{pmatrix} -\sin(\theta) & -\cos(\theta) & 0 \\ \cos(\theta) & -\sin(\theta) & 0 \\ 0 & 0 & 0 \end{pmatrix},$$

$$\hat{\hat{\mathbf{R}}}(\phi) = \begin{pmatrix} -\cos(\phi) & \sin(\phi) & 0 \\ -\sin(\phi) & -\cos(\phi) & 0 \\ 0 & 0 & 0 \end{pmatrix}. \tag{5.31}$$

This can also be written in terms of path axes as

$$\mathbf{R}'(\phi)\hat{\mathbf{R}}(\theta)\dot{\theta}\begin{pmatrix} V_b \\ 0 \\ 0 \end{pmatrix} = \mathbf{R}'(\phi)\hat{\mathbf{R}}(\phi)\dot{\phi}\begin{pmatrix} V_c \\ \dot{d} \\ 0 \end{pmatrix} + \begin{pmatrix} \dot{V_c} \\ \ddot{d} \\ 0 \end{pmatrix} + \mathbf{R}'(\phi)\hat{\hat{\mathbf{R}}}(\phi)\dot{\phi}^2\begin{pmatrix} 0 \\ d \\ 0 \end{pmatrix}$$

$$+ \mathbf{R}'(\phi)\hat{\mathbf{R}}(\phi)\ddot{\phi}\begin{pmatrix} 0 \\ d \\ 0 \end{pmatrix} + \mathbf{R}'(\phi)\hat{\mathbf{R}}(\phi)\dot{\phi}\begin{pmatrix} 0 \\ \dot{d} \\ 0 \end{pmatrix} \tag{5.32}$$

or

$$\hat{\mathbf{R}}_V\dot{\theta}\begin{pmatrix} V_b \\ 0 \\ 0 \end{pmatrix} = \mathbf{R}_\phi\dot{\phi}\begin{pmatrix} V_c \\ 2\dot{d} \\ 0 \end{pmatrix} + \begin{pmatrix} \dot{V_c} \\ \ddot{d} \\ 0 \end{pmatrix}$$

$$+ \hat{\mathbf{R}}_\phi\dot{\phi}^2\begin{pmatrix} 0 \\ d \\ 0 \end{pmatrix} + \mathbf{R}_\phi\ddot{\phi}\begin{pmatrix} 0 \\ d \\ 0 \end{pmatrix}, \tag{5.33}$$

with

$$\hat{\mathbf{R}}_\phi = \begin{pmatrix} \cos(\phi) & \sin(\phi) & 0 \\ -\sin(\phi) & \cos(\phi) & 0 \\ 0 & 0 & 1 \end{pmatrix}\begin{pmatrix} -\cos(\phi) & \sin(\phi) & 0 \\ -\sin(\phi) & -\cos(\phi) & 0 \\ 0 & 0 & 0 \end{pmatrix} = \begin{pmatrix} -1 & 0 & 0 \\ 0 & -1 & 0 \\ 0 & 0 & 0 \end{pmatrix},$$

$$\hat{\mathbf{R}}_V = \begin{pmatrix} \cos(\phi) & \sin(\phi) & 0 \\ -\sin(\phi) & \cos(\phi) & 0 \\ 0 & 0 & 1 \end{pmatrix} \begin{pmatrix} -\sin(\theta) & -\cos(\theta) & 0 \\ \cos(\theta) & -\sin(\theta) & 0 \\ 0 & 0 & 0 \end{pmatrix}$$

$$= \begin{pmatrix} -\cos(\phi)\sin(\theta) + \sin(\phi)\cos(\theta) & -\cos(\phi)\cos(\theta) - \sin(\phi)\sin(\theta) & 0 \\ \sin(\phi)\sin(\theta) + \cos(\phi)\cos(\theta) & \sin(\phi)\cos(\theta) - \cos(\phi)\sin(\theta) & 0 \\ 0 & 0 & 0 \end{pmatrix}$$

$$= \begin{pmatrix} -\sin(\theta - \phi) & -\cos(\theta - \phi) & 0 \\ \cos(\theta - \phi) & -\sin(\theta - \phi) & 0 \\ 0 & 0 & 0 \end{pmatrix}. \tag{5.34}$$

Hence

$$\begin{pmatrix} -\sin(\theta - \phi) & -\cos(\theta - \phi) & 0 \\ \cos(\theta - \phi) & -\sin(\theta - \phi) & 0 \\ 0 & 0 & 0 \end{pmatrix} \dot{\theta} \begin{pmatrix} V_b \\ 0 \\ 0 \end{pmatrix}$$

$$= \begin{pmatrix} 0 & -1 & 0 \\ 1 & 0 & 0 \\ 0 & 0 & 0 \end{pmatrix} \dot{\phi} \begin{pmatrix} V_c \\ 2\dot{d} \\ 0 \end{pmatrix} + \begin{pmatrix} \dot{V}_c \\ \ddot{d} \\ 0 \end{pmatrix}$$

$$+ \begin{pmatrix} -1 & 0 & 0 \\ 0 & -1 & 0 \\ 0 & 0 & 0 \end{pmatrix} \dot{\phi}^2 \begin{pmatrix} 0 \\ d \\ 0 \end{pmatrix} + \begin{pmatrix} 0 & -1 & 0 \\ 1 & 0 & 0 \\ 0 & 0 & 0 \end{pmatrix} \ddot{\phi} \begin{pmatrix} 0 \\ d \\ 0 \end{pmatrix}. \tag{5.35}$$

In terms of components, this yields

$$-\sin(\theta - \phi)\dot{\theta} V_b = -2\dot{d}\dot{\phi} + \dot{V}_c - \ddot{\phi}d,$$
$$\cos(\theta - \phi)\dot{\theta} V_b = \dot{\phi} V_c + \ddot{d} - \dot{\phi}^2 d. \tag{5.36}$$

Hence we obtain

$$\dot{V}_c = -\sin(\theta - \phi)\dot{\theta} V_b + 2\dot{d}\dot{\phi} + \ddot{\phi}d,$$
$$\ddot{d} = \cos(\theta - \phi)\dot{\theta} V_b - \dot{\phi} V_c + \dot{\phi}^2 d. \tag{5.37}$$

Substituting for $\dot{\phi}$, and noting that

$$\ddot{\phi} = \dot{\kappa} V_c + \kappa \dot{V}_c, \tag{5.38}$$

yields

$$\dot{V}_c = -\sin(\theta - \phi)\dot{\theta} V_b + 2\dot{d}\dot{\phi} + \ddot{\phi}d$$
$$= -\sin(\theta - \phi)\dot{\theta} V_b + 2\dot{d}\kappa V_c + d(\dot{\kappa} V_c + \kappa \dot{V}_c)$$

$$(1 - \kappa d)\dot{V}_c = -\sin(\theta - \phi)\dot{\theta}V_b + 2\dot{d}\kappa V_c + d\dot{\kappa}V_c$$

$$\dot{V}_c = \frac{-\sin(\theta - \phi)\dot{\theta}V_b + (2\dot{d}\kappa + d\dot{\kappa})V_c}{(1 - \kappa d)} \tag{5.39}$$

with

$$\ddot{d} = \cos(\theta - \phi)\dot{\theta}V_b - \dot{\phi}V_c + \dot{\phi}^2 d$$

$$= \cos(\theta - \phi)\dot{\theta}V_b - \kappa(1 - \kappa d)V_c^2. \tag{5.40}$$

Recall that the required characteristic equation is given by

$$\ddot{d} + 2\zeta\omega_n\dot{d} + \omega_n^2 d = 0. \tag{5.41}$$

Substituting for the first and second derivative of d yields

$$\cos(\theta - \phi)\dot{\theta}V_b - \kappa(1 - \kappa d)V_c^2 + 2\zeta\omega_n[V_b\sin(\theta - \phi)] + \omega_n^2 d$$

$$= \cos(\theta - \phi)\dot{\theta}V_b - \frac{\kappa V_b^2\cos(\theta - \phi)^2}{(1 - \kappa d)} + 2\zeta\omega_n[V_b\sin(\theta - \phi)] + \omega_n^2 d$$

$$= 0 \tag{5.42}$$

where

$$V_c = \frac{V_b\cos(\theta - \phi)}{(1 - \kappa d)}. \tag{5.43}$$

Solving for $\dot{\theta}$ gives

$$\dot{\theta} = \frac{\kappa V_b^2\cos(\theta - \phi)^2/(1 - \kappa d) - 2\zeta\omega_n V_b\sin(\theta - \phi) - \omega_n^2 d}{\cos(\theta - \phi)V_b}. \tag{5.44}$$

The demanded κ_d to the UAV is thus given by

$$\kappa_d = \frac{\dot{\theta}}{V_b}. \tag{5.45}$$

Hence we have

$$\kappa_d = \frac{\kappa V_b^2\cos(\theta - \phi)^2/(1 - \kappa d) - 2\zeta\omega_n V_b\sin(\theta - \phi) - \omega_n^2 d}{\cos(\theta - \phi)V_b^2}. \tag{5.46}$$

This gives the curvature demand for the UAV. Figure 5.4 shows the resulting trajectory, which follows the same Dubins path as that in the linear case. It also

Dubins Trajectory

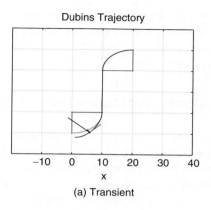

(a) Transient

Dubins Trajectory

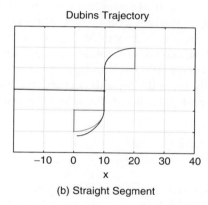

(b) Straight Segment

Dubins Trajectory

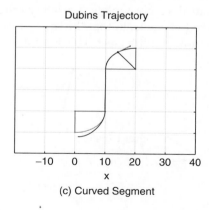

(c) Curved Segment

Figure 5.4   Nonlinear guidance trajectory

shows that the UAV follows both the straight and the curved segments of the Dubins path, which is an improvement over the linear guidance algorithm. The turn radius (which is the inverse of the curvature) is shown by the line normal to the UAV velocity vector.

5.4 Dynamic Obstacle Avoidance Guidance

If the obstacle is mobile, then use must be made of the velocity vector of the obstacle in order to predict when the obstacle and the UAV will occupy the same position some time in the future. This requires some form of estimator (usually a Kalman filter) to estimate the speed and direction of travel from sensor measurements made of the obstacle's position over time. Usually the moving obstacle is an aircraft of some form, so we will use this term for a moving obstacle from now on for a non-cooperating vehicle. It could also be another UAV, which will also manoeuvre to avoid a collision and so will be a cooperating vehicle; hence we will retain the UAV label for this case. The basic geometry of an aircraft and UAV collision course is shown in Figure 5.5.

The figure shows that a collision will occur if the ratio of the two sides of the *impact triangle* and the ratio of the velocity of the UAV to that of the aircraft are such that

$$\frac{d_\mathrm{u}}{d_\mathrm{a}} = \frac{v_\mathrm{u}}{v_\mathrm{a}}. \tag{5.47}$$

Then the two vehicles will reach the *impact point* at the same time. The collision avoidance algorithm must manoeuvre the UAV so that the closest approach to the aircraft does not violate the safety criterion. This criterion is the same as for a fixed obstacle, in that the safety circles surrounding the UAV and the aircraft must not intersect. This is shown in Figure 5.6. Hence

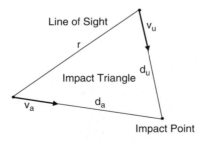

Figure 5.5   Basic geometry for collision of UAV and aircraft

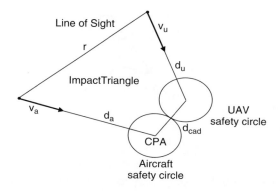

Figure 5.6 Basic geometry for collision avoidance of UAV and aircraft

the collision avoidance algorithm should detect the collision geometry and ensure that it at least meets the geometry in Figure 5.6.

The point at which the UAV is closest to the aircraft is known as the closest point of approach (CPA) and at this point the distance between the UAV and the aircraft is the closest approach distance d_{cad}. If the closest distance is greater than the safe distance (SD) d_{sd}, which is the sum of the radii of the UAV and aircraft safety circles, then the aircraft and the UAV are considered safe in following the current paths and no action is required. If, however, the closest approach distance is less than d_{sd}, then the path of the UAV must be modified to produce a closest approach distance greater than the SD. The collision avoidance (CA) algorithm is designed and developed based on the principle of proportional navigation (PN) and the airborne collision avoidance system known as TCAS.

Here we consider two approaches to resolve the collision conflict between UAVs and aircraft (Shin *et al.* 2008). We begin with collision avoidance with constant-speed manoeuvres for the UAVs for both single and multiple conflicts. Single conflicts involve one vehicle and a UAV, and multiple conflicts involve more than one vehicle with the UAV. This is then extended to the case where the speed of the UAV is variable.

In order to generate an autonomous collision avoidance algorithms for UAVs, it is first necessary to calculate the closest point of approach (CPA). To determine the CPA, consider the sightline geometry as shown in Figure 5.7. This shows that, if the relative velocity vector of the UAV v_r with respect to the aircraft is constructed, where

$$v_{ua} = v_u - v_a, \tag{5.48}$$

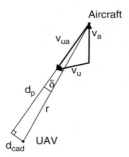

Figure 5.7   Sightline geometry for single UAV and aircraft

then the closest approach distance d_{cad} can be derived by placing the relative velocity vector v_{ua} on the aircraft and projecting it along a line towards the UAV. A line is then constructed normal to it that passes through the UAV. Hence, we have

$$d_{cad} = r \sin(\delta),$$
$$d_p = r \cos(\delta), \tag{5.49}$$

where d_p is the length of the relative velocity vector line. If d_{cad} is less than the minimum required distance d_{sd}, then a potential collision is detected. Hence, the condition for the existence of the collision is given by

$$d_{cad} < d_{sd}. \tag{5.50}$$

From this, the time to the CPA, which is used in the current TCAS algorithm, can be determined by

$$T_p = \frac{d_p}{v_{ua}}, \tag{5.51}$$

where v_{ua} is the magnitude of the relative velocity vector v_{ua}. The next consideration is how to change the geometry to avoid a potential collision. Hence, consider the manoeuvre required to change the relative velocity vector direction in order to produce a miss greater than d_{sd}. Figure 5.8 shows the geometry with a relative velocity vector and a circle radius d_{sd} imposed on the aircraft position.

As shown in the figure, the relative velocity vector v_{ua} should be positioned outside of the sector defined by the two tangent vectors from the aircraft

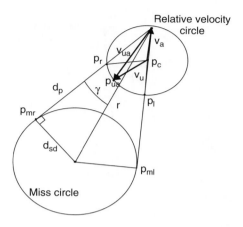

Figure 5.8 Sightline miss geometry for single UAV and aircraft

position to the tangent points p_{ml} and p_{mr} to achieve a miss distance greater than a miss of d_m. This is translated into the condition where the point p_{ua} should lie outside the sector on the relative velocity circle (RCA) defined by points p_r and p_l. The figure shows that the relative velocity vector v_{ua} is inside the sector, so indicating a potential collision. There are two control variables that can be used to change the geometry. The first is to keep the UAV velocity constant but change its direction. The second is to change the UAV velocity. We shall now consider each in turn.

5.4.1 UAV Direction Control

Using the assumption that the UAV speed is constant, the condition for collision avoidance can be achieved by changing the UAV heading. Figure 5.9 shows the required angle change to the UAV velocity vector to line it up with the nearest sector direction. The closest point on the relative velocity circle that will produce the collision avoidance condition is p_r. This will line up the relative velocity vector v_{ua} with the right-hand tangent line located at p_{mr}. Hence the desired UAV velocity vector is $\hat{v}_u$ for the single conflict resolution. Hence the relative velocity vector v_{ua} should be pointed outside of the sector $p_c p_r p_l$ to resolve the single vehicle conflict for constant UAV speed.

In order to compute the sector radii, consider the intersection of the tangent line with the UAV circle. As shown in Figure 5.10, using the cosine rule on triangle $p_c p_a p_l$, it is possible to derive the desired relative speed

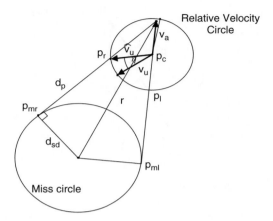

Figure 5.9   Geometry for single UAV and aircraft

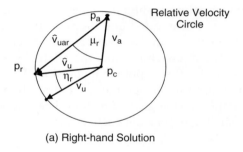

(a) Right-hand Solution

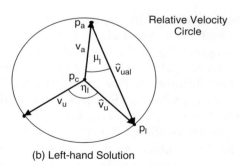

(b) Left-hand Solution

Figure 5.10   Chord intersection for sector definition

$\hat{v}_{ua}$ magnitude for the avoidance algorithm:

$$v_u^2 = \hat{v}_{ua}^2 + v_a^2 - 2\hat{v}_r v_a \cos(\mu_r)$$

$$0 = \hat{v}_{ua}^2 - 2v_a \cos(\mu_r)\hat{v}_{ua} + (v_a^2 - v_u^2)$$

$$\hat{v}_{ua} = v_a \cos(\mu) \pm \sqrt{v_u^2 - v_a^2 \sin^2(\mu)}. \tag{5.52}$$

Here μ represents the angle between the aircraft velocity vector v_a and $\hat{v}_{ua}$, v_u is the magnitude of the UAV velocity vector v_u, and v_a is the magnitude of the aircraft velocity vector v_a obtained from the estimator. The angle μ can be obtained by first calculating the tangent vector $\hat{v}_{ua}$ by rotating the sightline by angle γ, where

$$\gamma = \tan^{-1}\left(\frac{d_{sd}}{d_p}\right) \tag{5.53}$$

and

$$\mu = \cos^{-1}(v_a \cdot \hat{v}_{ua}), \tag{5.54}$$

where $(\cdot)$ represents the inner vector product of the two vectors. Finally, from equation (5.48), the desired UAV velocity vector $\hat{v}_u$ for single conflict resolution is given by

$$\hat{v}_u = \hat{v}_{ua} + v_a. \tag{5.55}$$

This approach can be extended to obstacle avoidance on both sides of the obstacle by computing the angle to turn anticlockwise instead of clockwise. The left or anticlockwise solution is also shown in Figure 5.10.

The simplest choice for calculation of the desired direction of rotation can be made by determining the closest point on the relative velocity circle. A simple algorithm that takes the UAV velocity vector out of the collision sector will be to define a turn rate proportional to the angular error η between the UAV velocity vector and the sector lines defined by either d_l or d_r. Turning towards the sector line d_r will make a clockwise turn, and turning towards the sector line d_l will make an anticlockwise turn. For the case shown in Figure 5.9, a clockwise turn will give a quicker response, as the UAV velocity vector travels a shorter distance around the circle. Turning clockwise will also give a monotonically increasing miss, whereas turning anticlockwise will initially reduce the miss distance and momentarily produce conditions for a collision before turning away again. Hence, if the guidance algorithm takes the UAV velocity v_u towards the closest sector vector, then the algorithm will produce a monotonically increasing miss distance. Other factors that will influence the turn direction for the avoidance guidance algorithm could be to obey rules of the air and mimic TCAS, or to allow more efficient following of an existing path after the avoidance manoeuvre.

The stability of the algorithm for a single conflict resolution can be determined by use of a simple Lyapunov candidate function V of the form

$$V = \tfrac{1}{2}\eta_e^2, \tag{5.56}$$

where

$$\eta_e = \min(\eta_l, \eta_r), \tag{5.57}$$

and where η_r and η_l represent the angle rotation required to achieve the velocity vectors for clockwise rotation $\hat{v}_{ur}$ and anticlockwise rotation $\hat{v}_{ul}$, respectively, as shown in Figure 5.10. To establish stability, the time derivative of V must be negative definite. The time derivative of V is given by

$$\frac{dV}{dt} = \dot{\eta}_e\eta_e. \tag{5.58}$$

Hence, for stability, we require an avoidance guidance command of the UAV that satisfies the following equation:

$$\dot{\eta}_e\eta_e \leq 0. \tag{5.59}$$

Now, for the case of fixed UAV speed, the circle radius in Figure 5.10 will remain unchanged. So, the angular rate of the error η_e is given by the difference between the angular rate of the desired UAV velocity vector $\hat{v}_u$ and the actual UAV velocity vector v_u, which can be set by the guidance algorithm. In order to calculate the angular rate $\dot{\hat{e}}$ (either right or left), consider the relationship between the desired velocity vector $\hat{v}_u$ (either right or left) and the relative velocity vector $\hat{v}_{ua}$ for single conflict resolution. The relative velocity vector from equation (5.48) gives

$$\hat{v}_u = \hat{v}_{ua} + v_a,$$
$$v_u\hat{t}_u = \hat{v}_{ua}\hat{t}_{ua} + v_a t_a, \tag{5.60}$$
$$\hat{t}_u = \alpha\hat{t}_{ua} + \beta t_a,$$

where $\alpha = \hat{v}_{ua}/v_u$, $\beta = v_a/v_u$, and t_u, t_{ua} and t_a are unit vectors. Figure 5.11 shows the vector sum defined in equation (5.60).

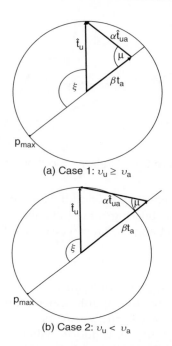

(a) Case 1: $v_u \geq v_a$

(b) Case 2: $v_u < v_a$

Figure 5.11 Velocity geometry. Reprinted with permission of the American Institute of Aeronautics and Astronautics

The figure shows that two solutions are possible depending on the ratio of v_u and v_a. The UAV guidance algorithm must generate a turn rate of v_u sufficient to converge onto the required direction $\hat{v}_u$. This can be achieved by determining the maximum rate of change of the vector $\hat{v}_u$ as the geometry changes. The geometry will change due to the motion of the UAV and aircraft as they approach one another. Noting that the aircraft is considered to be moving in a straight line with constant speed, and the UAV is flying at a fixed speed, the vector t_a and the variable α are fixed. Hence the solution vector $\hat{t}_{ua}$ and the variable β will vary with geometry, with the locus of the end point located on the circle in Figure 5.11. From the figure, the maximum rate of change of the desired velocity vector $\hat{v}$ occurs at point p_{max}, as shown in the figure. Hence, we have

$$\dot{\xi}_p = \alpha \dot{\mu}. \tag{5.61}$$

The rate of rotation of $\hat{t}_{ua}$ can be derived by reference to Figures 5.8 and 5.9. From Figure 5.9, it is assumed that the UAV can measure the properties of

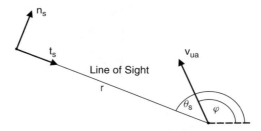

Figure 5.12 Relative position geometry

the line of sight between the UAV and the aircraft. So the UAV can measure the range, bearing and bearing rate of change of the aircraft. Hence, from Figure 5.8, the rate of change of $\hat{t}_{ua}$ can be derived as

$$\dot{\mu} = \dot{\zeta}_s + \dot{\gamma}, \tag{5.62}$$

where $\dot{\zeta}_s$ is the line-of-sight bearing rate.

In order to calculate the two angular rates $\dot{\zeta}_s$ and $\dot{\gamma}$, consider the sightline and express the relative velocity vector v_{ua} in terms of a basis vector set orientated along the sightline, t_s and n_s, where t_s is the basis vector along the sightline and n_s is the basis vector normal to the sightline, as shown in Figure 5.12. So, the relative position of the UAV to the aircraft can be written in the form

$$r = rt_s + 0n_s. \tag{5.63}$$

Differentiating with respect to time yields

$$\dot{r} = \dot{r}t_s + r\dot{t}_s$$
$$= \dot{r}t_s + r\dot{\zeta}_s n_s. \tag{5.64}$$

Now, the relative velocity v_{ua} can also be written in sightline basis vectors, to give

$$v_{ua} = -v_{ua}\cos(\varphi)t_s - v_{ua}\sin(\varphi)n_s. \tag{5.65}$$

Equating terms gives

$$\dot{r} = -v_r\cos(\varphi),$$
$$\dot{\zeta}_s = -\frac{v_r}{r}\sin(\varphi). \tag{5.66}$$

From Figure 5.7, $\sin(\gamma)$ and $\cos(\gamma)$ can be written in the form

$$\sin(\gamma) = \frac{d_m}{r},$$

$$\cos(\gamma) = \frac{\sqrt{r^2 - d_m^2}}{r}. \tag{5.67}$$

The angular rate of γ is obtained by differentiating $\sin(\gamma)$ with respect to time, to give

$$\dot{\gamma}\cos(\gamma) = -\frac{d_m}{r^2}\dot{r}$$

$$= \frac{d_m}{r^2} v_{ua}\cos(\varphi)$$

$$\dot{\gamma} = \frac{d_m}{r\sqrt{r^2 - d_m^2}} v_{ua}\cos(\varphi). \tag{5.68}$$

Substituting this and equation (5.66) into equation (5.62) yields

$$\dot{\mu} = \frac{v_{ua}}{r}\sin(\varphi) + \frac{d_m}{r\sqrt{r^2 - d_m^2}} v_{ua}\cos(\varphi), \tag{5.69}$$

which simplifies to

$$\dot{\theta}_l = \frac{v_r}{\sqrt{r^2 - d_m^2}}\sin(\gamma - \phi). \tag{5.70}$$

Substituting this into equation (5.61) gives

$$\dot{\xi}_p = \frac{a v_{ua}}{\sqrt{r^2 - d_m^2}}\sin(\gamma - \varphi). \tag{5.71}$$

In order to select a UAV turn rate that is greater than the solution rate of change, the maximum value of $\dot{\xi}_p$ is needed. This is given by

$$\max(v_{ua}) = (v_a + v_u),$$

$$\max(\alpha) = \frac{(v_a + v_u)}{v_u},$$

$$\max(\dot{\xi}_p) = \frac{(v_a + v_u)^2}{v_u\sqrt{r^2 - d_m^2}}. \tag{5.72}$$

Note that the angular rate $\dot{\xi}_r$ can be derived in the same way with $\dot{\xi}_l$. From this equation and equation (5.58), a UAV tangent vector control for the avoidance algorithm is proposed as

$$\dot{\xi}_u = \frac{(v_a + v_u)^2}{v_u\sqrt{r^2 - d_m^2}} \text{sign}(\eta_e) + K\eta_e, \qquad (5.73)$$

$$K > 0, \qquad \text{sign}(\eta_e) = |\eta_e|/\eta_e,$$

to give

$$\frac{dV}{dt} = \left[\dot{\eta}_{(l \text{ or } r)}\eta_e - \frac{(v_a + v_u)^2}{v_u\sqrt{r^2 - d_m^2}} \text{sign}(\eta_e)\eta_e \right] - K\eta_e^2 \leq 0. \qquad (5.74)$$

5.4.2 Multiple Conflict Resolution

Multiple conflicts are defined as the possibility of more than two collisions with other vehicles. The avoidance algorithm is extended to resolve multiple conflicts of the non-cooperating UAVs and aircraft. For simplicity, one UAV and two aircraft will be used to illustrate the extension. However, the proposed algorithm can be easily extended to general multiple conflict resolution. We consider the case of a UAV and two aircraft on a collision course simultaneously, as shown in Figure 5.13.

In order to determine the closest points of approach, the sightlines between the aircraft and the UAV are considered as in the case of single conflict

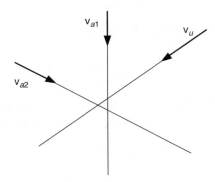

Figure 5.13   Collision condition for single UAV and two aircraft

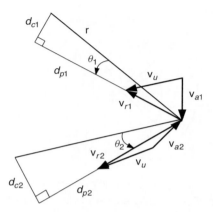

Figure 5.14 Sightline miss geometry for single UAV and two aircraft

determination. As shown in Figure 5.14, if the relative velocity vectors of the
UAV v_{ua} with respect to the aircraft are constructed, where

$$v_{ua1} = v_u - v_{a1},$$

$$v_{ua2} = v_u - v_{a2}, \tag{5.75}$$

then the distances of the closest points of approach d_{c1} and d_{c2} can be
calculated by using

$$d_{c1} = r\sin(\theta_1),$$

$$d_{c2} = r\sin(\theta_2). \tag{5.76}$$

If there is a possibility of multiple conflicts, the distances of the closest points
of approach are smaller than the minimum required distance d_m at the same
time. Hence, the condition for multiple conflicts is defined as

$$d_{c1} < d_m \quad \text{and} \quad d_{c2} < d_m. \tag{5.77}$$

As depicted in Figure 5.15, if there is the possibility of multiple conflicts, the
collision sectors are always overlapping. In order to resolve multiple conflicts
using the single conflict resolution concept, it is necessary to redefine the
collision sector for multiple conflict resolution. The sector δ can be defined

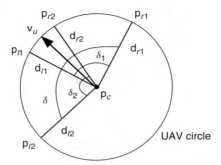

Figure 5.15 Chord intersection for multiple conflict resolution. Reprinted with permission of the American Institute of Aeronautics and Astronautics

from the vectors d_{r1} and d_{l2} by joining the centre point p_c to the points p_{r1} and p_{l2}.

The stability of the multiple conflict resolution algorithm can be determined by a simple Lyapunov candidate function L of the form

$$L = \tfrac{1}{2}\eta_e^2,$$

$$\eta_{el} = \eta_l - \eta_u,$$

$$\eta_{er} = \eta_r - \eta_u,$$

$$\eta_e = \min(\eta_{el}, \eta_{er}), \qquad (5.78)$$

where

$$\eta_l = \max(\eta_{l1}, \eta_{l2}),$$

$$\eta_r = \min(\eta_{r1}, \eta_{r2}). \qquad (5.79)$$

Then, the angular rate η_u for the multiple conflict resolution is derived as

$$\dot{\eta}_u = \lambda \, \mathrm{sign}(\eta_e) + K\eta_e, \qquad (5.80)$$

$$K > 0,$$

to give

$$\frac{dL}{dt} = [\dot{\eta}_{(l\ or\ r)}\eta_e - \lambda \, \mathrm{sign}(\eta_e)\eta_e] - K\eta_e^2 \le 0 \qquad (5.81)$$

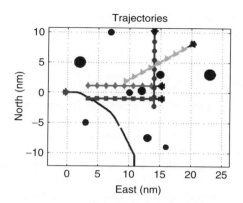

Figure 5.16   Avoidance algorithm trajectories for multiple aircraft. Reprinted with permission of the American Institute of Aeronautics and Astronautics

where

$$\lambda = \min\left(\frac{(v_{a1} + v_u)^2}{v_u\sqrt{r_1^2 - d_m^2}}, \frac{(v_{a2} + v_u)^2}{v_u\sqrt{r_2^2 - d_m^2}}\right). \tag{5.82}$$

Figure 5.16 shows the algorithm in action, in that the avoidance algorithm is used to avoid a collision with four aircraft approaching from different directions.

References

Caravita, L., Tsourdos, A., Aouf, N., White, B. A. and Silson, P. 2007. Control strategies applied to waypoint navigation and obstacle avoidance guidance. *Advances in Control and Optimization of Dynamical Systems, ACODS'2007*, 1–2 February. Indian Institute of Science, Bangalore.

Fossen, T. I. 1994. *Guidance and Control of Ocean Vehicles*. John Wiley & Sons, Ltd, Chichester (reprinted 1999).

Kaminer, I., Pascoal, A., Hallberg, E. and Silvestre, C. 1998. Trajectory tracking for autonomous vehicles: an integrated approach to guidance and control. *Proc. AIAA Journal of Guidance, Control, and Dynamics*, **21**(1), 29–38.

Kim, D. H. and Oh, J. H. 1999. Nonlinear tracking control of trailer systems using the Lyapunov direct method. *Journal of Robotics Systems*, **16**, 1–8.

Micaelli, A. and Samson, C. 1992. 3D path following and time varying feedback stabilization of a wheeled robot. *Proc. Int. Conf. on Control, Automation, Robotics and Vision, ICARCV92*, Singapore. RO-13.1.

Micaelli, A. and Samson, C. 1993. Trajectory tracking for unicycle-type and two-steering-wheels mobile robots. Technical Report no. 2097, INRIA, Sophia-Antipolis.

Nelson, D. R., Barber, B. and McLain, T. W. 2007. Vector field path following for miniature air vehicles. *IEEE Transactions on Robotics*, **23**(3), 519–529.

Park, S., Deyst, J. and How, J. 2004. A new nonlinear guidance logic for trajectory tracking. *Proc. AIAA Guidance, Navigation, and Control Conf.*, August. AIAA-2004-4900.

Pascoal, A., Silvestre, C. and Oliveira, P. 2006. Vehicle and mission control of single and multiple autonomous marine robots. *Advances in Unmanned Marine Vehicles* (eds Roberts, G. N. and Sutton, R.), IET Control Engineering Series, vol. 69, pp. 353–386. Institution of Engineering and Technology.

Shin, H. Tsourdos, A., White, B. A., Shanmugavel, M. and Tahk, M., UAV Conflict Detection and Resolution for static and dynamic obstacles, *AIAA Guidance, Navigation, Control Conf. and Exhibit, Honolulu, Hawaii, 18–21, August, 2008*, AIAA 2008-6521.

6

Path Planning
for Multiple UAVs

Simultaneous arrival at a specified location or locations is one of the missions that needs cooperation between UAVs. The mission objective is that a group of UAVs have to reach the destination(s) at the same time. The purpose of this mission may be to rendezvous for data exchange or to detect an object for tracking purposes. This problem is studied in McLain and Beard (2000) and McLain *et al.* (2001) using a global route planner, followed by refinement of the routes to produce flyable paths. Simultaneous arrival is achieved by producing paths of equal length, assuming that the UAVs are all flying at the same speed. An analogy of stretching a chain of springs connecting the waypoints is used to produce equal-length paths.

Consider N UAVs leaving the base at time t_{base} and which have to reach the target destination at the time t_{target}, where $t_{base} < t_{target}$. The base and target may be connected through a set of waypoints as shown in Figure 6.1. For simplicity, this figure shows only the base and the target locations. The environment may well have obstacles, which may require path modification to avoid collision (this topic is dealt with in Chapter 4). The poses of each UAV at the base are known and the poses at the target location are predefined to maximize the effectiveness of the sensor on each UAV. The UAVs are assumed to have equal kinematic and dynamic capabilities and are flying at the same speed in free space. Each UAV is assumed to lie at the centre of two concentric spheres, which lie at the centre of mass of the UAVs. The inner sphere is the safety sphere with radius R_s, while the outer one is the

Cooperative Path Planning of Unmanned Aerial Vehicles
Antonios Tsourdos, Brian White and Madhavan Shanmugavel
© 2011 John Wiley & Sons, Ltd

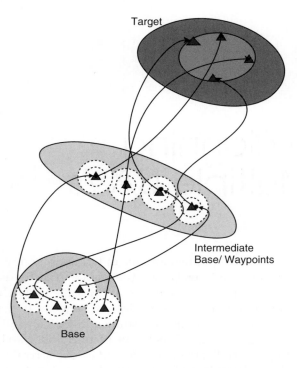

Figure 6.1 UAV safety and communication range spheres. Reprinted with permission of the American Institute of Aeronautics and Astronautics, Reproduced from IEEE

communication sphere of radius R_r, which represents the transmitter range such that $R_r > R_s \geq 1/\kappa_{max}$. Equivalently, the paths can be modelled as tubes with radii R_s around the paths, as shown in Figure 6.2. The generic version of this cooperative path planning can be stated as producing flight paths $\{r_1(q), r_2(q), \ldots, r_N(q)\}$ for UAVs $\{UAV1, UAV2, \ldots, UAVN\}$ such that, for the ith and jth UAVs,

$$\text{objective:} \quad h_{f,i} = h_{f,j}, \tag{6.1a}$$

$$\text{constraints:} \quad \amalg = \begin{cases} |v_i| = |v_j|, \\ \amalg_\kappa : \quad \kappa \leq \kappa_{max}, \\ \amalg_\tau : \quad \tau \leq \tau_{max}, \\ \amalg_{safe} : \quad r_i(q) \cap r_j(q) = \varnothing, \quad i \neq j, \\ R_r > R_s \geq 1/\kappa_{max}. \end{cases} \tag{6.1b}$$

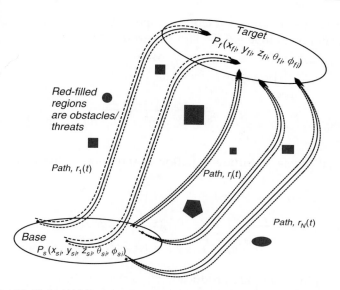

Figure 6.2   Multiple UAVs scenario: $r_i(t)$ are flight paths, (x, y, z, θ, ψ) are poses. Suffix i represents the ith UAV or path. Shaded regions are obstacles/threats

6.1 Problem Formulation

Consider path planning for a single UAV from the base to the target location. The path planner produces a path $r(q)$ connecting the start pose $P_s(x_s, y_s, z_s, \theta_s, \psi_s)$ and the finish pose $P_f(x_f, y_f, z_f, \theta_f, \psi_f)$, as defined in Chapter 1:

$$P_s(x_s, y_s, z_s, \theta_s, \psi_s) \xrightarrow{r(q)} P_f(x_f, y_f, z_f, \theta_f, \psi_f). \tag{6.2}$$

Extending equation (6.2) to account for N UAVs such that each UAV passes through n_p number of positions/poses gives

$$P_{s,i,j-1}(x_{s,i,j-1}, y_{s,i,j-1}, z_{s,i,j-1}, \theta_{s,i,j-1}, \psi_{s,i,j-1})$$
$$\xrightarrow{r_{i,j-1}(q),\ \sqcup_{safe},\ \sqcup_{length}} P_{f,i,j}(x_{f,i,j}, y_{f,i,j}, z_{f,i,j}, \theta_{f,i,j}, \psi_{f,i,j}), \tag{6.3}$$
$$|\kappa_i(q)| < \kappa_{max},\ |\tau_i(q)| < \tau_{max},\ i = 1, \ldots, N,\ j = 2, \ldots, n_p.$$

Here τ is the torsion, κ is the curvature, κ_{max} is the maximum-curvature bound, τ_{max} is the maximum-torsion bound, and $\sqcup_{safe}$ and $\sqcup_{length}$ are the constraints on safety and path length, respectively, as defined in Chapter 1. The constraint on path length $h(q)$ for the group of UAVs is

$$\underset{\text{length}}{\coprod} = \min \int_{s_1}^{s_2} |s(q)| \, dq \tag{6.4}$$

$$\text{where} \quad h(q) = \int_{q_1}^{q_2} \sqrt{\dot{x}(q)^2 + \dot{y}(q)^2 + \dot{z}(q)^2} \, dq, \quad q \in [q_1, q_2]. \tag{6.5}$$

For flights at constant altitude, equation (6.4) reduces to the 2D form

$$P_{s,i,j-1}(x_{s,i,j-1}, y_{s,i,j-1}, \theta_{s,i,j-1})$$

$$\xrightarrow{r_{i,j-1}(q), \, \coprod_{\text{safe}}, \, \coprod_{\text{length}}} P_{f,i,j}(x_{f,i,j}, y_{f,i,j}, \theta_{f,i,j}), \tag{6.6}$$

$$|\kappa_i(q)| < \kappa_{\max}, \quad i = 1, \ldots, N, \quad j = 2, \ldots, n_p.$$

Once the initial paths are produced, it is necessary to ensure that the paths are safe to fly and also that the UAVs accomplish their mission by arriving simultaneously at their destinations. Here a question arises: how and when to plan. This depends on the available information about the environment and location of the UAVs. One way is to integrate the path generation with the planning. This approach is similar to the route planning by optimization methods discussed in section 1.9. However, this approach needs further refinement to produce flyable paths and is also computationally intensive. Another possibility is to separate the two processes: path generation and planning. Now two possibilities arise: when to plan – either before or after the path generation. This question is addressed in section 6.2.

Mathematically, a path can be characterized by a curve. Therefore, the path planning can be considered as that of a geometric evolution of a curve. The flight path $r(q)$ in equation (6.4) has to meet the maximum-curvature bound $\kappa_{\max}$, maximum-torsion bound $\tau_{\max}$ and the constraints $\coprod$. The values of $\kappa_{\max}$ and $\tau_{\max}$ define the kinematic limits of the UAV. Also, these are the only two parameters that completely determine a curve in space (Kreyszig 1991; Lipschutz 1969). In two dimensions, only the curvature is required to determine the curve. A curve satisfying the curvature constraints imposed by the dynamics of the UAV is called a flyable path. A feasible path is thus defined as both flyable (meets kinematic and dynamic constraints) and safe to fly (no collisions).

The flight path $r(q)$ is either a single polynomial curve or a composite curve, as discussed in Chapters 2 and 3. Such a path is useful in predicting the position and attitude of the UAVs over the whole mission before the UAVs fly the paths. Also, it helps the path planner to consider the kinematic limits at the early phase of path planning.

6.2 Simultaneous Arrival

The solution is required to accomplish the mission of simultaneous arrival of a group of UAVs by satisfying the various constraints defined in the previous section of this chapter. The main constraints are the maximum bounds on the curvature and collision avoidance. The solution consists of three phases. The first phase produces flyable paths, the second phase then modifies the resulting paths to produce feasible paths, where the flyable paths now meet the safety constraints. Paths of equal length are then produced in the third phase. The proposed path planner is shown in Figure 6.3.

Depending on the speed constraints of the UAVs, simultaneous arrival can be achieved in two ways: (i) by producing paths equal in length for constant-speed UAVs, or (ii) by producing paths of unequal lengths for variable-speed UAVs. Here, only constant-speed UAVs are considered and, as all UAVs are assumed to be flying at the same constant speed, producing paths of equal length ensures their simultaneous arrival. Considering simplicity in representation, only two waypoints are considered, but multiple waypoints can be handled if required. Hence path planning requires the solution to the equation:

$$P_{si}(x_{si}, y_{si}, z_{si}, \theta_{si}, \psi_{si}) \xrightarrow{\amalg_{\text{safe}}, \amalg_{\text{length}}, r_i(q)} P_{fi}(x_{fi}, y_{fi}, z_{fi}, \theta_{fi}, \psi_{fi}), \qquad (6.7)$$

$$|\kappa_i(q)| < \kappa_{i,\max}, \ |\tau_i(q)| < \tau_{i,\max}, \ h_i(q) = h_j(q),$$

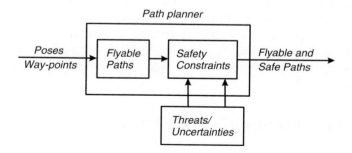

Figure 6.3   Path planner for flyable paths. Reprinted with permission of Elsevier

where $h_i(q)$ and $h_j(q)$ are the path lengths of the ith and jth UAV. The path length $h(q)$ of the path $r(q) = \{x(q), y(q), z(q)\}$ is given by

$$h(q) = \int_{q_1}^{q_2} \sqrt{\dot{x}(q)^2 + \dot{y}(q)^2 + \dot{z}(q)^2}\, dq, \qquad q \in [q_1, q_2], \tag{6.8}$$

where $\dot{x}(q) = dx/dq$, $\dot{y}(q) = dy/dq$ and $\dot{z}(q) = dz/dq$ are hodographs.

Equation (6.8) can be solved by optimization techniques, but this will be computationally intensive and the resultant path may not be optimal. Considering these difficulties, an alternative approach is required. This consists of breaking the problem into manageable subproblems. Hence, the solution is divided into three phases: (i) producing flyable paths, (ii) producing feasible (safe and flyable) paths, and (iii) producing paths of equal length.

6.3 Phase I: Producing Flyable Paths

In the first phase, flyable paths are produced by connecting the start and finish poses of each UAV. The design of flyable paths are discussed in Chapters 2 and 3. These paths are labelled initial paths. The design of flyable paths can be written as

$$P_{si}(x_{si}, y_{si}, z_{si}, \theta_{si}, \psi_{si}) \xrightarrow{r_i(q)} P_{fi}(x_{fi}, y_{fi}, z_{fi}, \theta_{fi}, \psi_{fi}), \tag{6.9}$$

$$|\kappa_i(q)| < \kappa_{i,\max} \; |\tau_i(q)| < \tau_{i,\max}.$$

Three types of paths can be used: (i) Dubins paths, (ii) Pythagorean hodograph (PH) paths or (iii) clothoid paths. The Dubins and clothoid are composite paths that are piecewise continuous, while the PH is a single path and is produced by interpolation. As we have seen in the previous chapters, the Dubins path provides a path with up to the maximum-curvature bound. For a more continuous path that a fixed-wing UAV can follow more easily, two types of paths can be used: (i) Pythagorean hodograph (PH) paths, and (ii) clothoid paths.

6.4 Phase II: Producing Feasible Paths

In the second phase, the flyable paths are tuned to meet the safety constraints. Meeting these constraints ensures collision avoidance and sufficient space to

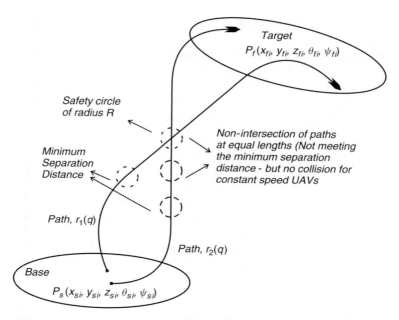

Figure 6.4 Safety constraints for collision avoidance. Reprinted with permission of Elsevier, and ASME

manoeuvre away from any obstacle. A feasible path is both flyable and safe. Flyable paths satisfy the principal maximum-curvature constraint. However, this path does not guarantee safety until the safety constraints are imposed on the flyable paths. Two safety constraints that are defined to avoid collision between UAVs and other obstacles are: (i) minimum separation distance, and (ii) non-collision of paths. These two constraints are shown in Figure 6.4. Each UAV is assumed to have a safety radius R_s, located at the centre of mass. This safety radius is less than the sensor range but greater than the minimum radius of curvature.

6.4.1 Minimum Separation Distance

Consider first the case where each of the flight paths do not intersect. The distance between any two flight paths should be greater than a threshold value given by the minimum separation distance, d_{sep}. Here the threshold is taken as twice the safety radius R_{safe}. This can be interpreted as testing for non-intersection of the flight paths, and for offset curves or tubes for meeting the safety margin. This is shown in Figure 6.4, where the paths $r_1(q)$ and $r_2(q)$ are separated by d_{sep}.

The implementation of minimum separation distance is as follows. Referring to Figure 6.4, safety circles are centred on points along each path. If these safety circles of any two paths overlap, a collision could occur. Representing the distance between two paths, $r_k(q)$ and $r_m(q)$, by $d_{sep,k,m}$, the constraint of minimum separation distance can be written as

$$d_{sep,k,m} > R_{s,k} + R_{s,m}, \tag{6.10}$$

where $R_{s,k}$, and $R_{s,m}$ are the safety radii of the kth and mth paths, respectively.

Hence, the minimum separation distance d_{sep} between any two UAV paths should be equal to or greater than the sum of the corresponding radii of their safety circles. The separation between two UAVs is measured by calculating the Euclidean distance between two points on two different paths. The separation distance between the kth path and the mth path at a particular length or time is given by

$$d_{sep,k,m} = \sqrt{(z_m - z_k)^2 + (y_m - y_k)^2 + (x_m - x_k)^2}, \tag{6.11}$$

where (x_k, y_k, z_k) is the point on the kth path and (x_m, y_m, z_m) is the point on the mth path. For the 2D case, we have

$$d_{sep,k,m} = \sqrt{(y_m - y_k)^2 + (x_m - x_k)^2}, \tag{6.12}$$

and for homogenous UAVs with identical safety distances, equation (6.10) reduces to

$$d_{sep,k,m} > 2R_{safe}. \tag{6.13}$$

6.4.2 Non-Intersection Paths

In many cases the computed paths may have intersections. This will violate the constraints dealt with in the previous section. In such cases, the 'minimum separation distance' test will give a false result. If paths intersect, it does not necessarily mean that a collision occurs. The UAVs will only collide if the UAVs reach the intersection point at the same time. The safety circles will intersect if the UAVs are within $d_{sep,k,m}$ as they approach the intersection point.

Referring to Figure 6.4, the paths $r_k(q)$ and $r_m(q)$ avoid collision if the difference between the lengths of the paths to the intersection point d_{int} meet the following condition:

$$d_{int,k,m} = |s_{int,k} - s_{int,m}| > d_{sep} > (R_{s,k} + R_{s,m}). \tag{6.14}$$

Here $s_{int,k}$ and $s_{int,m}$ are the lengths of the paths $r_k(q)$ and $r_m(q)$, respectively, from their starting points to the intersection point. If the paths meet the safety constraints defined in equations (6.10) and (6.14), there is no need to re-plan the paths. Otherwise, the paths are re-planned by varying the curvature or by flying through intermediate waypoints and/or poses.

If the flyable path meets the minimum path separation condition in equation (6.10), the paths are safe to fly and there is no need to re-plan the path. If this condition fails, then condition (6.14) is tested for intersection collision of the paths. In the event of failure of both conditions, then re-planning is required. For a group of N UAVs, taking n UAVs at a time, the number of safety tests T_{safe} to be carried out is

$$T_{safe} = 2\frac{N!}{n!(N-n)!}. \tag{6.15}$$

6.4.3 Offset Curves

Incrementing or moving a circle or sphere along the flight path and then constructing the evolute curve produces either two offset curves for the 2D case or a tube for the 3D case. In the 2D case, the minimum separation distance can be represented by a set of evolute offset paths. The offset paths $r_d(q)$ at an offset distance R_{safe} are given by

$$r_d(q) = r(q) \pm R_{safe}n(q), \tag{6.16}$$

where $n(q)$ is the normal vector to $r(q)$, and the offset distance is equal to twice the safety radius.

Similarly, for a 3D path, the canal surface or tube $r_s(q)$, with radius equal to the safety radius, is given by

$$r_s(q) = r(q) + R_{safe}[\, n(q) \;\; b(q)\,]\begin{bmatrix} \cos(\theta) \\ \sin(\theta) \end{bmatrix}, \tag{6.17}$$

where $n(q)$ is the normal vector and $b(q)$ is the binormal vector, which can be analytically computed from curve parametrization.

6.5 Phase III: Equalizing Path Lengths

Simultaneous arrival on target requires the equalization of all path lengths for a group of UAVs flying at the same speed. Though the second phase of the path planner produces feasible paths (flyable and safe), they are not usually of equal length. A reference path length is required to provide a target for the path length modification. Paths can be lengthened by changing curvature constraints (see Chapter 2). The lengths of all paths are increased to that of the reference path by suitable manipulation of the curvature constraints.

If the reference path corresponding to the path of maximum length is given by r_{ref} and its length is given by s_{ref}, then

$$r_{ref} = r(s_{ref}), \qquad i = 1, \ldots, N. \qquad (6.18)$$

The value of the curvature κ is modified to satisfy

$$h_{ref} - \{h_i\} = 0, \qquad i = 1, \ldots, N - 1, \qquad (6.19)$$

where h_i is the length of the ith path and N is the number of UAVs. Note that the reference path is not included in the modification process.

6.6 Multiple Path Algorithm

A simple algorithm for producing a set of safe paths for a group of UAVs is as follows:

(i) Produce flyable paths for each UAV to meet the curvature constraint.
(ii) Test the paths to meet the safety constraints for collision avoidance.
(iii) If paths fail to meet the safety constraint, modify the paths either by reducing the curvature or by producing intermediate waypoints.
(iv) Calculate the length of the paths and select the reference path length as the maximum of the path lengths.
(v) Increase the length of the shorter paths to that of the reference path without violating safety constraints.

The first step in the algorithm is straightforward, as the flyable path is generated by connecting the given poses or waypoints. The computation time increases with the number of UAVs, as the number of safety constraint tests will increase with the UAVs (see equation (6.15)). The computation speed can be accelerated by using a global path planner to produce waypoints and/or poses or by producing intermediate waypoints (see Chapter 4).

6.7 Algorithm Application for Multiple UAVs

The algorithm is applied to path planning using all of the path planning routines described in Chapters 2 to 4.

6.7.1 2D Dubins Paths

To illustrate the use of the path planning algorithms, consider planning a set of paths for five UAVs. For simplicity, only two waypoints and/or poses are considered, and these are designated the base and target locations. The above algorithm is then used to determine a set of paths with equal lengths. Once the longest path between the start and finish locations is designated as the reference path, the remaining paths can then be made to have the same length by using a simple binomial search over the range of permitted curvatures for each path. The largest radius arc and hence the smallest curvature κ_{min} is derived from equation (2.27), which results in a path with no straight segment. This gives a lower bound on the path curvature, with the upper bound κ_{max} predetermined by the maximum curvature that the UAV can follow. Hence we have

$$\kappa_{min} < \kappa < \kappa_{max}. \tag{6.20}$$

A simple Euler search algorithm will then iterate onto a curvature that gives the required path length for each of the other Dubins paths in the set. The minimum turning radius is set at 1.2 units.

A safety radius is next set for each UAV, which will ensure that no two paths approach closer than the safety radius. This then ensures that the UAVs will not get too close to each other and thus avoid collision. This topic is further developed in Chapter 4. Note that the closest approach is more complex than ensuring that the paths do not encroach on the safety radius, as paths can be close provided that the relative difference in length is not less than the safety radius. The radius of the safety circle is 2.5 units. The orientations at the finish poses are taken as free variables so that the UAVs can approach the target from either direction along the orientation direction. The shortest path is thus chosen from the set of eight available paths for each UAV. These paths are $r_i (i = 1, \ldots, 5)$ for each of the four UAVs and are the

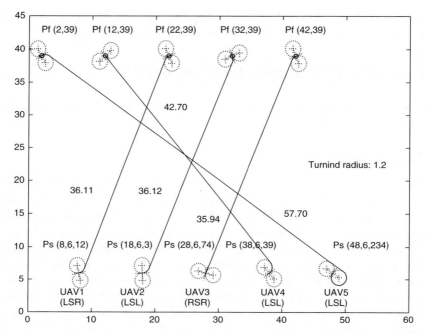

Figure 6.5 UAV shortest flyable paths – Dubins 2D. Reprinted with permission of the American Institute of Aeronautics and Astronautics, and ASME

initial paths as shown in Figure 6.5. The five paths have different lengths and are respectively LSR, LSL, RSR, LSL and LSL Dubins paths.

Using equation (6.18), the reference path is found to be r_5. Hence the lengths of $r_j (j = 1, \ldots, 4)$ have to be increased to that of r_5. However, solving equation (2.13) or (2.33) for a given length does not result in a unique solution and may result in a complex solution. Hence, the Euler search algorithm is used to calculate the optimal radii to produce paths of equal lengths. The resulting optimal radii of $r_j (j = 1, \ldots, 4)$ are {9.82, 4.38, 9.86, 14.42} units, respectively. These radii are then used to produce the final paths of equal length.

Figure 6.6 shows the final equal-length paths of the UAVs. From the figure, it can be observed that the route or trace of the final paths are different from that of the initial paths. The paths have changed from {LSR, LSL, RSR, LSL} to {LSL, LSL, RSL, LSL}, where the routes r_1 and r_3 are changed. This is because the planner takes the path of minimum length from the set of available paths. If the first available path could not meet the reference length requirement, the planner picks the next shortest path.

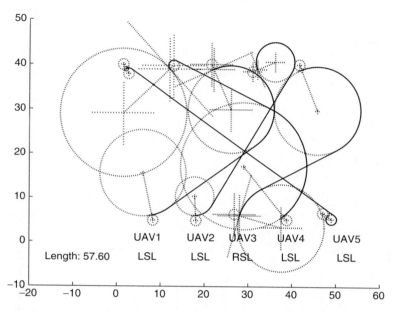

Figure 6.6 Paths of equal length – Dubins 2D. Reprinted with permission of the American Institute of Aeronautics and Astronautics, and ASME

The paths of equal length are tested for safety using equations (6.13) and (6.14). This is done by finding the intersection of any two paths at a time. The resulting intersection can be of a line with a line, a line with a circular arc, or an arc with an arc. Thus, there are 10 combinations possible for five paths (see Chapter 4). Figure 6.7 shows the flight paths of UAV1 with other UAVs in the group. The top left corner shows that the paths r_1 and r_2 intersect at two points. The minimum distance between the paths is 5.1 units, which just meets the minimum separation requirement. (Note that each safety circle is of radius 2.5 units, giving a safety distance of 5 units.) As the UAVs are assumed to fly at constant speed, and have the same start time, then, at each increment of path length over time, there is no collision between the UAVs. The same argument is applied to the other three set of paths: r_1 with r_3, r_1 with r_4, and r_1 with r_5.

The first three parts of Figure 6.8 show the interactions of flight path r_2 with r_3, r_4 and r_5, respectively. The first two of these meet the minimum separation constraint, while the third pair, r_2 with r_5, does not. However, this pair has non-intersecting paths and hence eliminates the possibility of collision between UAV2 and UAV5. The last part of Figure 6.8 shows the

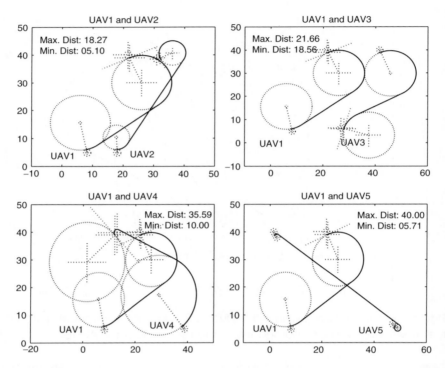

Figure 6.7 Separation distance for paths of first four combinations –
Dubins 2D. Reprinted with permission of the American Institute of Aeronautics
and Astronautics, and ASME

interaction between r_3 and r_4, and the first part of Figure 6.9 shows the
interaction between r_3 and r_5. These two paths meet the minimum separation
condition and hence are safe. The interaction between r_4 and r_5 shown in
the last part of Figure 6.9 also meets the minimum separation, and hence
provides a safe path. Thus the interaction between the flight paths of all five
UAVs meet both the minimum safety distance and non-intersection of paths
at equal lengths. This provides safe paths for all UAVs. Figure 6.6 shows the
final paths for simultaneous arrival on target.

6.7.2 2D Clothoid Paths

The same path planning situation as that solved by the Dubins paths in
section 6.7.1 is calculated with a group of three UAVs, flying at a constant
speed and at constant altitude. The UAVs are UAV1, UAV2 and UAV3. The

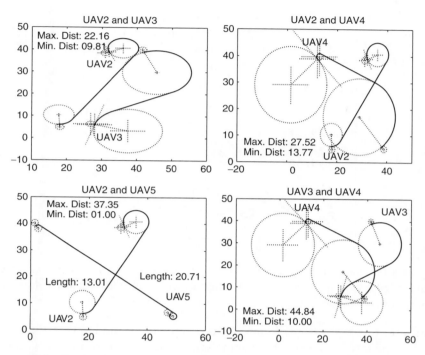

Figure 6.8 Separation distance for paths of second four combinations – Dubins 2D. Reprinted with permission of the American Institute of Aeronautics and Astronautics, and ASME

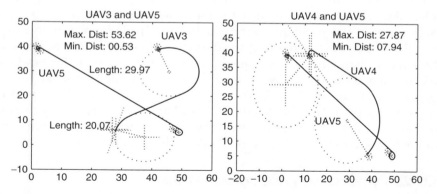

Figure 6.9 Separation distance for paths of last two combinations – Dubins 2D. Reprinted with permission of the American Institute of Aeronautics and Astronautics, and ASME

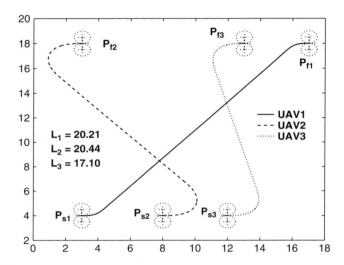

Figure 6.10 Initial flyable paths of UAVs – clothoid 2D. All UAVs have different path lengths and the paths are intersecting with one another. Reprinted with permission of Elsevier

start and finish positions, P_s and P_f, respectively, of the UAVs are defined and all the UAVs leave the base at the same time. The maximum curvatures, κ_{max}, of the UAVs is set as $\pm\frac{1}{4}$. Figure 6.10 shows the initial paths generated with the maximum-curvature bound of the UAVs. These are all flyable paths, as they meet the curvature constraint.

The figure shows that the initial flight paths intersect. Hence, safety constraints have to be evaluated for the paths. The paths are tested for minimum separation distance using equation (6.13) and for non-intersection of paths at any path length using equation (6.14). In this case, the paths meet the safety constraints. Next, the length of the shorter paths are increased to match that of the reference using equation (6.18). The path length of UAV1 is longer than that of UAV2 and UAV3, hence the path of UAV1 is designated as the reference path. The path lengths of UAV2 and UAV3 are then increased to that of UAV1 by decreasing the curvature of their clothoid segments. Figure 6.11 shows the final paths of UAVs, which pass the constraint tests.

6.8 2D Pythagorean Hodograph Paths

Three UAVs, UAV1, UAV2 and UAV3, flying at constant speed and at constant altitude are considered. The initial and final poses of the UAVs are

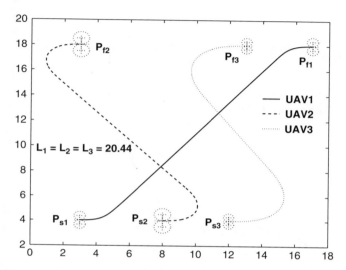

Figure 6.11 Final paths of UAVs (paths of equal lengths) – clothoid 2D. Reprinted with permission of Elsevier

fixed, as before. The maximum curvature of the UAVs is κ_{max} and is set to $\pm\frac{1}{3}$, and all the UAVs leave the base at the same time. Figure 6.12 shows the curvature-optimized paths. The start and finish points are shown with the tangent circles, which also define the maximum curvature of the UAVs. The safety conditions are tested for the three paths using equations (6.13) and (6.14). Taking two UAVs at a time, there are a total of six safe flight paths to be tested (according to equation (6.15)).

The path lengths of two of the UAVs are increased to the length of the reference path, which again is the longest path based on the initial PH paths. The lengths of the remaining PH paths are then increased by increasing the length of the boundary tangent vectors using equations (2.95) and (2.96). This condition is implemented algorithmically as

$$\text{find } \kappa \text{ such that } h_i - h_{\text{ref}} = 0, \qquad i = 1, \ldots, N - 1, \qquad (6.21)$$

where h is the length of the path.

Figure 6.13 shows the paths with their offset paths. The central path (solid line) is the flight path. The dashed path on either side of the flight path shows offset paths with safety circles. The offset paths are generated at a distance of the radius of the safety circle. Two important points to be considered from the figures are: the paths have curvature continuity, thus providing

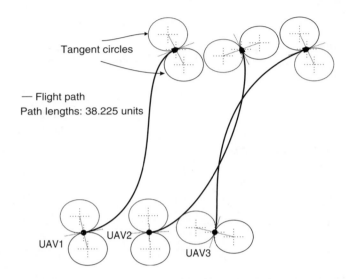

Figure 6.12　PH paths of equal lengths. Reprinted with permission of ASME

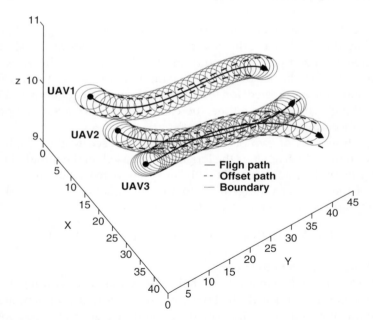

Figure 6.13　PH Paths of UAVs, equal lengths elevated at constant altitude. Reprinted with permission of ASME

smoothness, and each path has a different route or trace. The path r_1 does not intersect with the other paths. The other paths r_2 and r_3 intersect each other at two points. The points of intersection are found by an iterative search. The difference in the lengths of these paths from their initial points are then found to be 8.381 and 7.321, respectively. The values are greater than the diameter of the safety circle and so this ensures safe flight paths. The lengths of r_1 and r_3 are then increased to that of the reference path r_2 by using equation (2.96). The maximum and minimum curvatures of the paths are $(0.1142, -0.3000)$, $(0.3263, 0.002)$ and $(0.2718, 0.0055)$, respectively.

6.9 3D Dubins Paths

Three UAVs are considered for simulation. The minimum turning radius is chosen as 5. The initial and final poses are as follows:

UAV1: $\quad P_s(0.0, 0.0, 0.0, 10.0, 0.0) \xrightarrow{r_1(\text{flight path 1})} P_f(51.0, 18.0, 51.0, -10.0, 30.0)$,

UAV2: $\quad P_s(4.0, 7.0, 5.0, -10.0, 0.0) \xrightarrow{r_2(\text{flight path 2})} P_f(61.0, 18.0, 51.0, 70.0, 30.0)$,

UAV3: $\quad P_s(15.0, 0.0, 5.0, -28.0, 0.0) \xrightarrow{r_3(\text{flight path 3})} P_f(61.0, 45.0, 51.0, 10.0, 30.0)$.

The radius R_s of the safety sphere is 3. The minimum separation distance, d_{sep}, is 6 units. The flyable path for each UAV is generated using the principle explained in Chapter 3. The length of the flight path of each UAV is calculated using equation (3.21). Figure 6.14 shows the flyable paths of each UAV. The flight paths 1, 2 and 3, respectively, correspond to UAV1, UAV2 and UAV3. The length of each path is 76.27, 79.57 and 79.91, respectively. The length and track of each path are different, and also they intersect one another. Hence they do not meet the minimum separation distance. The path r_3 is the reference path found by equation (6.18). Therefore, the path lengths of r_1 and r_2 have to be increased to that of r_3. The turning radii of UAV1 and UAV2 are increased to equalize their length with that of UAV3.

Figure 6.19 (later) shows the intersection of flight paths. The separation distance d_{sep} between any two paths is equal to zero, as they intersect with each other. Hence the second safety constraint, non-intersection at equal length, has to tested for the paths. The values of $d_{int,i,j}$ for each possible pair of paths are: (i) $d_{int,1,2} = 5.47$, (ii) $d_{int,2,3} = 9.60$, and (iii) $d_{int,1,3} = 13.10$. As the value of $d_{int,1,2}$ is less than $(R_{s1} + R_{s2})$ (where R_{si} is the safety radius of the ith UAV), the trace of r_1 is changed by increasing the curvature to meet the

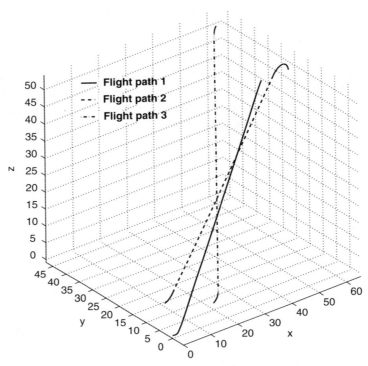

Figure 6.14 Flyable paths of UAVs – Dubins 3D. Reprinted with permission of the American Institute of Aeronautics and Astronautics

safety constraint and is followed by equalizing its length with that of UAV3, s_{ref}. As the curvatures determine a path completely, and there is no direct relation between the path length and turning radius, an iterative method is sought to find the optimal curvature of the paths to make them equal in length and also for meeting the safety constraint.

Therefore, the curvature of the path r_1 is reduced into 0.0759 so that $d_{int,1,2}$ becomes 6.863, and the length r_1 is equal to that of r_{ref}. Also, the length of path r_2 is increased to that of r_{ref} by reducing the curvature to 0.1074. A further test shows that the values of $d_{int,2,3}$ and $d_{int,1,3}$ are 6.541 and 15.48, which are greater than the sum of their respective safety radii. Hence they meet the safety requirements and result in paths of equal length, as shown in Figure 6.15. Figures 6.16, 6.17 and 6.18 each show two flight paths that are equal in length. The intersection of the paths in each plane is shown in Figure 6.19.

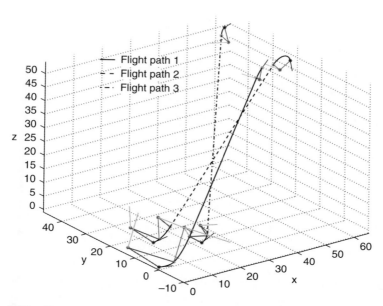

Figure 6.15 Paths of equal lengths – Dubins 3D. Reprinted with permission of the American Institute of Aeronautics and Astronautics

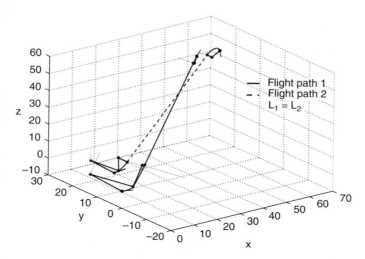

Figure 6.16 Paths of equal length (UAV1 and UAV2) – Dubins 3D. Reprinted with permission of the American Institute of Aeronautics and Astronautics

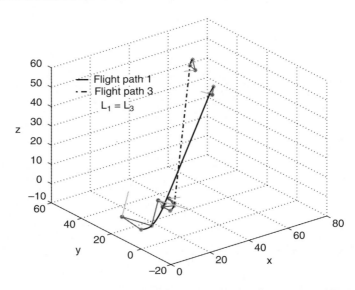

Figure 6.17 Paths of equal length (UAV1 and UAV3) – Dubins 3D. Reprinted with permission of the American Institute of Aeronautics and Astronautics

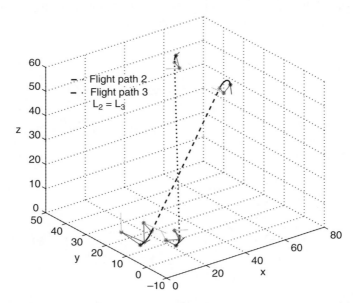

Figure 6.18 Paths of equal length (UAV2 and UAV3) – Dubins 3D. Reprinted with permission of the American Institute of Aeronautics and Astronautics

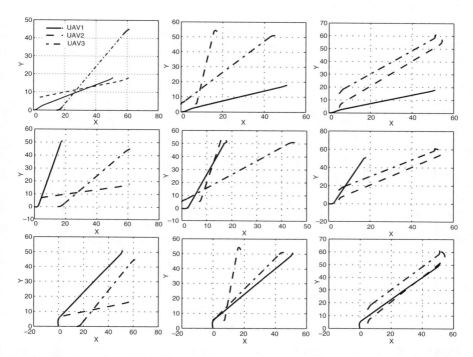

Figure 6.19 Flight path intersections. The intersections are given in each plane for each UAV. The intersections are calculated numerically. This is to avoid the possibility of complex points during the intersection between lines and circles. The following safety conditions are tested individually on all planes: (i) minimum separation distance, and (ii) non-intersection at equal lengths. Reprinted with permission of the American Institute of Aeronautics and Astronautics

6.10 3D Pythagorean Hodograph Paths

Two UAVs, UAV1 and UAV2, are considered for the simulation. The initial and final poses of the UAVs are predefined. The maximum curvatures, κ_{max} and τ_{max}, of the UAV are taken as $\pm\frac{1}{3}$. In contrast to the Dubins and clothoid paths, the PH path is a single tangent-continuous path obtained using Hermite interpolation (see section 3.3.1). The variations of the curvature and torque of these paths are shown in Figures 6.20 and 6.21. The peaks at the boundaries show that the paths do not meet the maximum-curvature and maximum-torque bounds.

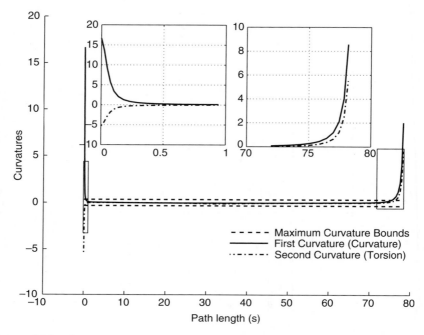

Figure 6.20 Curvature and torsion variations with respect to path length, tangent-continuous path – UAV1. The path does not meet the maximum-curvature bounds at the boundaries. Reprinted with permission of the American Institute of Aeronautics and Astronautics

The paths are optimized for curvature bounds by increasing the positive constants iteratively in equations (3.30c) and (3.30d) till the conditions (3.44) and (3.47) are met. The resulting flight paths are shown in Figure 6.22, which are initial paths. They are flyable by the UAVs. The corresponding curvature and torque variations of r_1 and r_2 with path length are shown in Figures 6.23 and 6.24. The path meets the maximum-curvature and maximum-torque bounds at all points of the path. The curvature and torque at the boundary points are now within the limits. The variation of these parameters at intermediate points is minimal compared with the variation in Figures 6.20 and 6.21.

From Figure 6.22, it can be seen that the paths do not meet the minimum separation distance and also their path lengths are not equal. Therefore, the paths have to be manipulated to produce paths of equal length. As the path

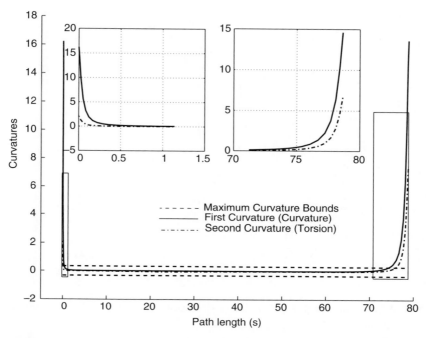

Figure 6.21 Curvature and torsion variations with respect to path length, tangent-continuous path – UAV2. The path does not meet the maximum-curvature bounds at the boundaries. Reprinted with permission of the American Institute of Aeronautics and Astronautics

length of UAV1 is longer than that of UAV2, the length of r_1 is the reference path r_{ref}. The length of UAV2 has to be increased to that of UAV1 by varying the curvature and torsion (see section 3.4) and also the safety constraints are to be met.

The modified path r_2 in Figure 6.25 shows that it intersects with r_2. However, they meet the constraint of non-intersection at equal length, $d_{int,i,j}$, so they are safe to fly. Figure 6.26 shows the variations of curvature and torsion of the path r_2 after modification. The tubes around the paths are produced by equation (6.17). The radius of each tube is equal to that of the safety sphere. As the calculation of the intersection between surfaces is computationally intensive, the tubes are produced after the paths meet the mission objective. Thus spatial PH paths of equal length for a group of UAVs achieving simultaneous arrival on target is accomplished.

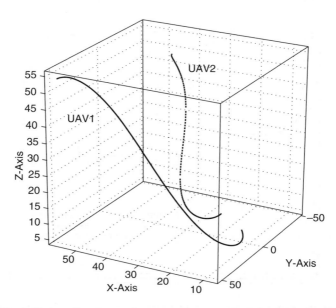

Figure 6.22 Initial paths (curvature-continuous and flyable) for UAV1 (solid line) and UAV2 (dotted)

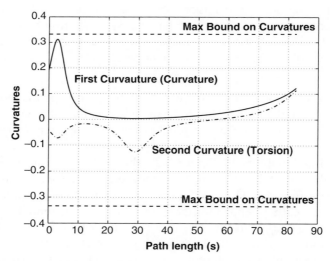

Figure 6.23 Curvature and torque variation of flyable path r_1 of UAV1 – PH 3D. Reprinted with permission of the American Institute of Aeronautics and Astronautics

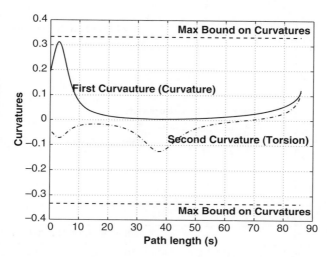

Figure 6.24 Curvature and torque variation of flyable path r_2 of UAV2 – PH 3D. Reprinted with permission of the American Institute of Aeronautics and Astronautics

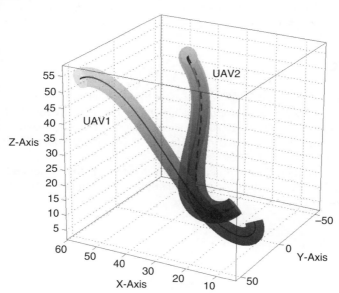

Figure 6.25 Final, feasible paths (solid) – UAV1 and UAV2 (dashed line is the initial path of UAV2). Reprinted with permission of the American Institute of Aeronautics and Astronautics

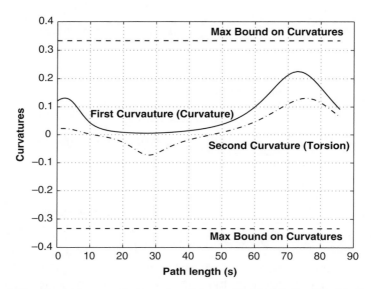

Figure 6.26 Curvature variation: feasible path of UAV2 – PH 3D (length is equal to that of UAV1). Reprinted with permission of the American Institute of Aeronautics and Astronautics

References

Kreyszig, E. 1991. *Differential Geometry*. Dover Publications.

Lipschutz, M. 1969. *Schaum's Outline of Differential Geometry*. McGraw-Hill.

McLain, T. and Beard, R. W. 2000. Trajectory planning for coordinated rendezvous of unmanned air vehicles. *AIAA Guidance, Navigation and Control Conf.*, Denver. AIAA-2000-4369, AAO-37126.

McLain, T., Chandler, P., Rasmussen, S. and Pachter, M. 2001. Cooperative control of UAV rendezvous. *Proc. American Control Conf.*, Arlington, VA, 25–27 June, pp. 2309–2314.

Appendix A

Differential Geometry

Differential geometry is the study of geometry using the principles of calculus. In general, a curve $r(q)$ is defined as a vector-valued function in $\Re^n$ space. The parameter q varies over a $\Re$ number line. Mathematically, this is a continuous mapping $r : I \to \Re^n$, where $I \in [a, b]$ and $q \in I$. For example, a curve $r(q)$ in 3D is represented as $r : I \to \Re^3$, where $r(q) = (x(q), y(q), z(q))$. Thus a curve $r(q)$ can be considered as a position vector in Euclidean space. The function $r(q)$ traces the curve as the parameter q varies. If the parameter q is time, the position vector will be given by a vector from the origin to the curve $(x(q), y(q), z(q))$ at time q. The velocity and acceleration can simply be calculated by taking the derivative of the curve, and their profiles can be drawn by substituting the values of q.

The geometric properties of the curve or path *per se* can be studied by unit speed parametrisation as follows. The arc length $h(q)$ of the curve $r(q)$ is

$$h(q) = \int_{s_1}^{s_2} \sqrt{\dot{x}^2 + \dot{y}^2 + \dot{z}^2} \, dq. \tag{A.1}$$

The unit speed parametrisation is such that the parametric speed $\dot{s} = ds/dq$ of the path is unity. This is an ideal concept. This is explained as follows. Consider a vehicle that starts moving at time q_1 and then stops at time q_2. The path length at time q_1 is h_1 and at time q_2 is h_2. A path of unit speed parametrisation has $(q_2 - q_1) = (h_2 - h_1)$. This means that the time travelled

---

Cooperative Path Planning of Unmanned Aerial Vehicles
Antonios Tsourdos, Brian White and Madhavan Shanmugavel
© 2011 John Wiley & Sons, Ltd

is equal to the distance travelled. Mathematically, this is

$$\left|\frac{dr}{dh}\right| = \frac{|dr/dq|}{|dh/dq|} = 1. \tag{A.2}$$

The physical significance of differential geometry of the curve is as follows. Taking q as time, the first derivative is the tangent vector and it defines velocity. The speed is given by the modulus of the velocity vector, and the direction of velocity (heading) is specified by the unit tangent vector, t. The second derivative is the acceleration vector, and this has two components, one along the tangent and other normal to the tangent. The tangential acceleration is given by the second derivative of the velocity vector and its direction is along the direction of the heading velocity. The direction of the normal acceleration is given by a unit normal vector, n, and its magnitude is equal to the centripetal acceleration given by $\kappa |v|^2$, where κ is the curvature and v is the velocity. Thus the curvature is proportional to the lateral acceleration and hence the lateral force induced while the vehicle is turning. Taking the path length as a parameter, the rate of change of the tangent vector with respect to the arc length defines the tangent vector.

The cross-product of the unit vectors t and n produces a third unit vector, called the binormal vector b, which is orthogonal to t and n. Thus the orthogonal triad (t, n, b) forms a moving frame on the curve. The plane spanned by the vectors t and n is the osculating plane. The vectors n and b form the normal plane, and the vectors b and t form the rectifying plane. These three planes are orthogonal to each other. A continuous sequence of this triad represents the orientation of the curve in space. The curvature and torsion (κ and τ) completely specify a path in space. Thus we have:

$$\text{unit tangent vector,} \qquad t = \frac{\dot{r}(q)}{|\dot{r}(q)|}, \tag{A.3}$$

$$\text{unit binormal vector,} \qquad b = \frac{\dot{r}(q) \times \ddot{r}(q)}{|\dot{r}(q) \times \ddot{r}(q)|}, \tag{A.4}$$

$$\text{unit normal vector,} \qquad n = b \times t. \tag{A.5}$$

The curvature profile at a point P is defined by the relation

$$\kappa = \frac{d\epsilon}{dh}, \tag{A.6}$$

where h is the path length and ϵ is the angle subtended by the tangent with the x axis. But,

$$\frac{d\epsilon}{dh} = \frac{d\epsilon/dq}{dh/dq}.$$

Hence, equation (A.6) becomes

$$\omega = v\kappa, \tag{A.7}$$

where $\omega = d\epsilon/dq$ is the angular velocity, $v = dh/dq$ is the linear velocity and q is the path parameter.

A.1 Frenet–Serret Equations

At every point on the curve, we can fix a local frame formed by the tangent, normal and binormal orthonormal vectors. Such a frame is called the Frenet–Serret (FS) frame (Figure A.1). The FS equations describe the

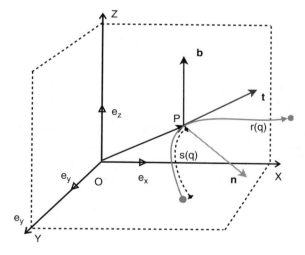

Figure A.1   Frenet–Serret frame $\{t, n, b\}$, in which t is the unit tangent, n is the unit normal and b is the unit binormal. On the diagram, $r(q)$ is the path, P is the position vector of a point on the path, $\{e_x, e_y, e_z\}$ are the unit vectors and $h(q)$ is the path length

rate of change of the curve with respect to the change of arc length. The FS equations are as follows:

$$t' = \kappa(q)n, \tag{A.8}$$

$$n' = -\kappa(q)t + \tau(q)b, \tag{A.9}$$

$$b' = -\tau(q)n. \tag{A.10}$$

In matrix form, this becomes

$$\begin{pmatrix} t' \\ n' \\ b' \end{pmatrix} = \begin{pmatrix} 0 & \kappa(q) & 0 \\ -\kappa(q) & 0 & \tau(q) \\ 0 & -\tau(q) & 0 \end{pmatrix} \begin{pmatrix} t \\ n \\ b \end{pmatrix}, \tag{A.11}$$

where the prime represents the derivative with respect to the path variable q and

$$\text{curvature,} \quad \kappa(q) = \|r'(q) \times r''(q)\|, \tag{A.12}$$

$$\text{torsion,} \quad \tau(q) = \frac{[r'(q) \cdot r''(q) \times r'''(q)]}{\kappa^2(q)}. \tag{A.13}$$

The time rate of change of the FS vectors in matrix form is

$$\begin{pmatrix} \dot{t} \\ \dot{n} \\ \dot{b} \end{pmatrix} = \dot{q} \begin{pmatrix} 0 & \kappa(t) & 0 \\ -\kappa(t) & 0 & \tau(t) \\ 0 & -\tau(t) & 0 \end{pmatrix} \begin{pmatrix} t \\ n \\ b \end{pmatrix}, \tag{A.14}$$

where $\dot{q} = dq/dt$ is the speed (parametric speed) and q is the path parameter. Thus we obtain

$$\text{curvature,} \quad \kappa(t) = \frac{\|\dot{r}(t) \times \ddot{r}(t)\|}{\|\dot{r}(t)\|^3}, \tag{A.15}$$

$$\text{torsion,} \quad \tau(t) = \frac{\dot{r}(t) \cdot \ddot{r}(t) \times \dddot{r}(t)}{\|\dot{r}(t) \times \ddot{r}(t)\|^2}. \tag{A.16}$$

A.2 Importance of Curvature and Torsion

Mathematically, a flyable path is a regular curve that captures both the geometric (locus of points) and kinematic (motion) aspects. A regular curve r is a mapping $r : [a, b] \to \mathbf{R}$ at least three times continuously differentiable, $r \in \mathbf{C}^3$ and satisfying the regularity condition $dr/dq \neq 0$ for all $q \in [a, b]$. Regularity means that the point moving along the curve is not allowed to stop, a natural requirement for fixed-wing UAVs. However, considering the kinematic constraints, it is important for the path to have curvature continuity.

By the principles of differential geometry (Kreyszig 1991; Lipschutz 1969), the curvature and torsion are fundamental properties of a path, by which a curve is completely determined in space. In two dimensions, curvature alone is enough. Apart from the geometric insights, these two properties play an important role in the mechanics of a moving vehicle. The physical significance of these properties are that the curvature is proportional to the lateral acceleration and is measured by the rate of change of the tangent vector, while the torsion is proportional to the angular momentum and is measured by the rate of change of the tangent plane:

$$\kappa(q) = \frac{\dot{r} \times \ddot{r}}{|\dot{r}|^3}, \tag{A.17}$$

$$\tau(q) = \frac{\|\dot{r}, \ddot{r}, \dddot{r}\|}{|\dot{r} \times \ddot{r}|^3}. \tag{A.18}$$

From equation (A.18), the curvature and torsion, respectively, are functions of the first two and three derivatives of the path. Hence, it is necessary to have a path of minimal order sufficient to satisfy curvature constraints and additional flexibility to negotiate obstacles.

A.3 Motion and Frames

The design of the Dubins path using analytic geometry is as simple and easy to understand as the Euclidean space is familiar to us. However, for an autonomous vehicle, it would be appropriate to use frames to describe the motion. A curve can be studied by assigning a frame at each point on it. The curve evolves with the rate of change of these frames (O'Neill 1967). The Frenet–Serret frame (FS) is one such frame, shown in Figure A.2. This frame constitutes tangent (t), normal (n) and binormal (b) vectors, which together form a trihedron on every point of the path. The advantage of the frame is that the rate of change of the trihedron varies with the frame itself with the given curvatures of the path:

$$t = \frac{\dot{r}}{|\dot{r}|}, \qquad n = \frac{\dot{r} \times \ddot{r}}{|\dot{r} \times \ddot{r}|}, \qquad b = t \times n, \tag{A.19}$$

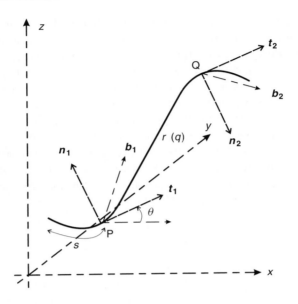

Figure A.2 Frenet–Serret frame on a 3D curve

where the derivatives are with respect to the path parameter q. The rate of change of these vectors (and hence the frame) is a function of two parameters, curvature and torsion:

$$\begin{pmatrix} \dot{t} \\ \dot{n} \\ \dot{b} \end{pmatrix} = v(q) \begin{pmatrix} 0 & \kappa & 0 \\ -\kappa & 0 & \tau \\ 0 & -\tau & 0 \end{pmatrix} \begin{pmatrix} t \\ n \\ b \end{pmatrix}, \quad\quad\quad (A.20)$$

where v is the velocity.

From equation (A.20), it can be seen that, for a given curvature and torsion, the evolution of the FS frame with time is the frame itself.

Kinematics can be best represented using differential geometry. Differential geometry enables the motion along the curve to be understood rather than representing motion with respect to some fixed frame. For example, taking time q as the path parameter, the path $r(q)$ represents the equation of motion of the vehicle along the path with time. Also, the motion can be expressed in terms of a moving trihedron along the curve. This moving trihedron is purely a function of the intrinsic properties of the path, namely (i) curvature and (ii) torsion, as shown in Figure A.2.

The Frenet–Serret (FS) frame forms a basis at each point on the curve. Hence, the curve can be studied and generated by transformation of these bases. From Figure A.2, the curve $r(q)$ is generated by transforming the frame $F_1(t_1, n_1, b_1)$ at P to a new frame $F_2(t_2, n_2, b_2)$ at Q by $F_2 = R_\epsilon F_1$, where R_ϵ is the rotation matrix with rotation angle ϵ.

References

Kreyszig, E. 1991. *Differential Geometry*. Dover Publications.
Lipschutz, M. 1969. *Schaum's Outline of Differential Geometry*. McGraw-Hill.
O'Neill, B. 1967. *Elementary Differential Geometry*. Academic Press.

Appendix B

Pythagorean Hodograph

Pythagoras's theorem for a right-angled triangle is well known from the principles of geometry. The theorem states that the square of the length of the hypotenuse of a right-angled triangle is equal to the sum of the squares of the lengths of the other two sides. For example, for a triangle with two sides a and b, and hypotenuse c, the theorem states that

$$c^2 = a^2 + b^2, \tag{B.1}$$

$$c = \sqrt{a^2 + b^2}. \tag{B.2}$$

If we extend equation (B.2) to polynomials $a(q)$, $b(q)$ and $c(q)$, it becomes

$$c(q)^2 = a(q)^2 + b(q)^2, \tag{B.3}$$

$$c(q) = \sqrt{a(q)^2 + b(q)^2}. \tag{B.4}$$

If the term inside the square root is a perfect square in the above two equations, the value of c or $c(q)$ would be an integer or a polynomial. Such a condition shows important properties, which provides the fundamental idea for the Pythagorean hodograph (PH). Here only a simplified version of the PH curve is given. A comprehensive treatment on the subject is given in (Farouki 2008). Here only a motivation of the subject is given.

Cooperative Path Planning of Unmanned Aerial Vehicles
Antonios Tsourdos, Brian White and Madhavan Shanmugavel
© 2011 John Wiley & Sons, Ltd

Consider a vehicle flying in a polynomial curve r. At any point on the curve, due to the turning, the vehicle will have a lateral velocity in addition to velocity in the tangential direction. That is, the total velocity v is decomposed into tangential velocity v_t along the tangent vector and lateral velocity v_n along the normal vector (see Appendix A):

$$v = \sqrt{v_t^2 + v_n^2},\tag{B.5}$$

$$\frac{dr}{dq} = \sqrt{\left(\frac{dr_t}{dq}\right)^2 + \left(\frac{dr_n}{dq}\right)^2}.\tag{B.6}$$

Here dr_t/dq is the velocity component along the tangent vector and dr_n/dq is the velocity component along the normal vector. Comparing equation (B.6) with equation (B.4) shows the similarities and possible extension of the theorem to physical applications.

B.1 Pythagorean Hodograph

Pythagorean hodographs are known for their rational properties. A rational curve is an irreducible curve whose parametrisation, for example $(x(q), y(q))$, would generate all points on the curve for the values of q, which is a parameter. Consider a path $r(q) = (x(q), y(q))$ with length $h(q)$:

$$h(q) = \int_{q_1}^{q_2} \sqrt{\dot{x}^2 + \dot{y}^2}\, dq,\tag{B.7}$$

where q is the path length variable, $\dot{x} = dx/dq$, $\dot{y} = dy/dq$ and $q \in [q_1, q_2]$. The parametric speed is $\dot{h} = dh/dq$. If $\dot{q} = \dot{r}$, this is called unit speed parametrisation. To calculate the path length exactly without any approximation, equation (B.7) should have a closed-form solution. This is not easy to obtain, even for simple polynomials, except for a straight line (Farouki and Sakkalis 1991).

The derivatives $\dot{x}$ and $\dot{y}$ are called hodographs. The path length is a function of the hodographs. To arrive at a simple solution without any approximation, the term inside the square-root term should be a square of some polynomial $\sigma(q)$. In other words, if a polynomial $\sigma(q)$ is selected such that

$$\sigma(q)^2 = x(q)^2 + y(q)^2,\tag{B.8}$$

then such formulations eliminate the approximation in calculation of the path length. Also, it results in useful rational properties of the path. Using the basic algebraic formulae $(a+b)^2 = a^2 + b^2 + 2ab$, and $(a^2 - b^2) = (a+b)$ $(a-b)$, the hodographs can be formulated using polynomials $u(q)$, $v(q)$ and $w(q)$ such that

$$\dot{x}(q) = [u(q)^2 - v(q)^2]w(q), \tag{B.9a}$$

$$\dot{y}(q) = 2u(q)v(q)w(q). \tag{B.9b}$$

This then implies that

$$\sigma(q) = [u(q)^2 + v(q)^2]w(q). \tag{B.10}$$

For the case where $w(q) = 1$ and the greatest common denominator $GCD(u(q), v(q)) = 1$, the resulting path is known as a regular PH path; that is, $GCD(\dot{x}, \dot{y}) = 1$ is a polynomial of odd degree.

Such a formulation produces the path length s and parametric speed $\dot{s}$ as polynomials. Also, the offset path, curvature κ and torsion all become rational:

$$h(q) = \int_0^1 w(q)[u(q)^2 + v(q)^2]\,\mathrm{d}q, \qquad q \in [0,1], \tag{B.11}$$

$$t = \frac{u(q)^2 - v(q)^2 2u(q)v(q)}{u(q)^2 + v(q)^2}, \tag{B.12}$$

$$n = \frac{2u(q)v(q)v(q)^2 - u(q)^2}{u(q)^2 + v(q)^2}, \tag{B.13}$$

$$\kappa = \frac{2[u(q)\dot{v}(q) - v(q)\dot{u}(q)]}{w(q)[u(q)^2 + v(q)^2]^2}. \tag{B.14}$$

References

Farouki, R. T. 2008. *Pythagorean-Hodograph Curves, Algebra and Geometry Inseparable.* Springer.

Farouki, R. T. and Sakkalis, T. 1991. Real rational curves are not "unit speed". *Computer Aided Geometric Design*, **8**, 151–157.

Index

acceleration, 176
  centripetal, 176
  lateral, 5, 176
  normal, 176
  tangential, 176
algorithm
  guidance, 119
angular rate calculation, 140
arc intersection, 94
autonomy
  architecture, 9
autopilot, 5, 119

bank angle, 7
bank-to-turn, 7
base, 149
basis vector, 140
bearing, 140
bearing rate, 140
binomial search, 157
binormal, 68, 176

carrot, 125
  distance, 126
cell decomposition, 17
closest approach distance, 134
closest point of approach (CPA), 133

clothoid, 19–20, 46, 160
collision, 154
collision avoidance, 8,
    81, 133
  algorithm, 142
  constant speed, 135
  stability, 138
collision detection, 85
  arc intersection, 94
  Dubins 2D, 85
    line intersection, 86
  line segment intersection, 90
collision geometry, 133
communication sphere, 148
configuration space, 13, 84
conflict resolution
  single, 137
conflicts
  multiple , 142
constraint
  curvature, 179
  dynamic, 4
  kinematic, 4–5
  path length, 150
  path planning, 3
  safety, 3, 7, 152
cooperative path planning, 7, 148

Cooperative Path Planning of Unmanned Aerial Vehicles
Antonios Tsourdos, Brian White and Madhavan Shanmugavel
© 2011 John Wiley & Sons, Ltd

curvature, 5, 45, 178–9
  clothoid, 46
  Dubins, 46
  yaw rate, 7
curvature continuity, 163
curve, 150, 175
  regular, 178

differential geometry, 175
  Frenet–Serret frame, 177
Dijkstra, 114
Dubins path, 19, 30, 157
  2D
    analytical geometry, 31
    differential geometry, 39
    path length, 39, 43
  3D, 165
    differential geometry, 67
    path length, 72
  CCC, 30
  CLC, 30
  condition for existence, 37

Euler search, 157–8
evolutionary algorithm, 19
external tangent, 33

feasible path, 152
filter
  Kalman, 132
flight path, 150
flyable path, 4, 153
frame
  Frenet–Serret, 68, 179
  trihedron, 179
Frenet–Serret frame, 177

guidance, 10, 120
  algorithm, 126, 139
  closest point of approach (CPA), 133
  direction control, 135
  linear, 124
  Lyapunov, 138
  nonlinear dynamic inversion, 126

  obstacle avoidance (dynamic), 132
guidance systems, 5

hodograph, 184

impact triangle, 132
initial path, 152
intermediate waypoint, 104, 157
internal tangent, 35
interpolation, 74, 152

Kalman filter, 132
kinematics, 180
kinematic limit, 150
kinematic model, 5

lead time, 81
line of sight, 140
line segment intersection, 90
linear guidance, 124
Lyapunov stability, 138

MILP, 18
minimum separation distance, 153
miss distance, 134, 137
mission, 10
multiple conflicts, 133

non-intersection of paths, 154
nonlinear dynamic inversion, 126
normal, 68, 176
number of tests, 155

obstacle avoidance
  2D, 82
  3D, 112
  clothoid, 107
  Dubins 2D, 107
  dynamic, 132
  impact point, 132
  impact triangle, 132
  intermediate waypoint algorithm, 104
  multiple conflicts, 142
  PH 2D, 110

safety circle, 132
safety circle avoidance, 104
static, 81
TCAS, 138
triangulation, 112
unmapped obstacles, 103
offset curve, 155
optimal control, 18
orientation, 3

parametric speed, 175
parametrisation
unit speed, 175
path
continuous curvature, 45
Dubins, 30
equal length, 151, 156
algorithm, 156
feasible, 150
flyable, 5, 152, 178
offset, 155
reference, 156
safe, 152
shortest, 30
path following, 119
carrot, 125
Dubins, 120
linear guidance, 124
nonlinear dynamic inversion, 126
obstacle avoidance (dynamic), 132
path length, 184
path planning, 2, 10
2D, 29
3D, 65
black box, 4
cooperative, 7
definition, 2
equation, 3
obstacle avoidance, 82
problem formulation, 149
pose, 2
potential field, 16
probabilistic method, 16

proportional navigation, 133
Pythagoras's theorem, 183
Pythagorean hodograph, 19, 184
2D, 56, 162
flyable path, 61
path length, 57
3D, 72, 169
flyable path, 74
path length, 73

range
transmitter, 148
reference path, 156
relative velocity circle, 135
road map, 13
roll rate, 7
route planning, 3

safe distance (SD), 133
safe path, 7, 152
safety circle, 132, 154
algorithm, 104
safety radius, 147, 153
safety sphere, 147
separation distance, 154
simultaneous arrival, 147
clothoid, 160
constant speed, 151
Dubins
2D, 157
3D, 165
PH
2D, 162
3D, 169
solution, 151–2
first phase, 152
second phase, 152
third phase, 156
variable speed, 151
stability
Lyapunov, 144

tangent, 68, 176
tangent circle, 31

target, 149
TCAS, 133
torsion, 179
    roll rate, 7
trajectory
    cubic spline, 16
trajectory tracking, 119
triangulation, 114
    Delaunay, 112
tube, 148, 155

UAV, 1, 7

velocity
    heading, 176
visibility graph, 14
Voronoi, 14

waypoint, 2

yaw rate, 7